Copyright © by Rand McNally & Co.

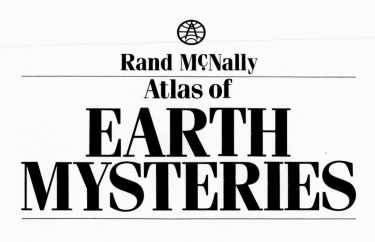

Rand McNally
Atlas of
EARTH
MYSTERIES

Consultant Editor PHILIP WHITFIELD

Atlas of
EARTH
MYSTERIES

Consultant Editor PHILIP WHITFIELD

Rand McNally

Chicago/New York/San Francisco

Rand McNally
Atlas of Earth Mysteries

Published by Rand McNally
in 1990 in the U.S.A.
Copyright © 1990 Marshall Editions
Developments Limited

Library of Congress Card Number: 89-43714
ISBN: 0528 83394-4

Typeset by Servis Filmsetting Limited,
Manchester, UK
Originated by Imago Publishing Ltd, Thame, UK
Printed and bound in West Germany
by Mohndruck Graphische Betriebe GMBH

A Marshall Edition
conceived, edited and designed by
Marshall Editions
170 Piccadilly
London W1V 9DD

Editor:	Gwen Rigby
Managing Editor:	Ruth Binney
Assistant Editor:	Sandy Shepherd
Art Editor:	Lynn Bowers
Research:	Jazz Wilson
Picture Research:	Richard Philpott
Production:	Barry Baker Janice Storr
Index:	Judith Beadle

Contents

Introduction

The Earth is filled with places that are permeated with the ether of mystery. Many have been created by humankind, but most intriguing and awe-inspiring are those natural locations and phenomena which are an integral part of the Earth itself.

Such mysterious sites and situations, although they are the subject of serious and intensive scientific investigation and research, retain their capacity to evoke wonder and stir the imagination. Science can go a long way toward providing convincing explanations, but many of these occurrences remain elusively enigmatic.

The pages that follow are a journey through the wonders of our planet, from the familiar to the bizarre. The mysteries encountered are many and various. They may be processes driven by forces infinitely removed from our everyday lives, such as the gravity that generates the ocean's tides. Or they may emerge from the massive workings of the Earth beneath our feet, through which the slow, but immeasurably powerful, jostlings of the continental plates give rise to volcanoes and earthquakes, geysers and giant waves. Then there are the climatic wonders—wonders with slow rhythms such as the mysterious fluctuations of El Niño, lightning strikes and tornadoes that are almost instantaneous, and the eccentric phenomena of creature storms and red rain.

On the loom of planetary activity whole landscapes are woven, replete with complexity and mystery. Here are specific mixes of geology, animal and plant life, such as the Great Rift Valley and the Sargasso Sea, which are effortlessly capable of invoking wonder in the beholder.

The mysterious forces of nature determine the patterning of our world map on the grand scale and in intimate detail. In an age when concern for the future of our planet is uppermost in people's minds, it has never been more important to divine its secrets and preserve its wondrous features for future generations.

The violent explosion of a volcano, bursting through the cold seas just south of Iceland on 14 November 1963, heralded one of the Earth's most dramatic and mysterious events in modern times: the birth of a new island. Surtsey has provided scientists with a unique opportunity to chart the colonization of sterile land by plants and animals, and to watch the subtle burgeoning of life on Earth.

Cosmic connections

Our world, the planet Earth, is full of wonders and mysteries. Perhaps the most remarkable is actually the least remarked upon: the extraordinary fact that we live out our lives on the surface of a thin film of solid rock floating on molten lava, spinning through space. As we dig our gardens, wash our teeth, feed babies, make love or commute to work, we are simultaneously careering through the vastness of interstellar space, stuck to the surface of a rocky planet and cocooned only by a thin strip of inhabitable atmosphere. What could be more mysterious than this mind-boggling incongruity—ordinary lives being lived out on a seemingly vulnerable planetary spaceship?

As you sit, apparently motionless, reading this sentence, the Earth's spin is whipping you around at some 800km/500mph, while the Earth's orbit around the Sun is imposing on you a speed along the orbit track of more than 80,000kmh/50,000mph. And you and the rest of the Solar System are majestically circling the Milky Way Galaxy! For our Sun with its attendant planets has no special place in the Universe. It is one among billions of others stars in a spiral galaxy—an island universe in its own right. And that home galaxy, which the ancients first named the Milky Way because its innumerable stars merged into a single milky streak in the night sky, is in no way unique. It is one of uncountable myriads of other galaxies in the observable Universe.

We hardly ever think of it, yet we cannot truly escape the cosmic context of our lives, for the cosmological background intrudes into workaday life in more ways than we realize—many of them explicable in only the most general terms by modern scientific theories. The Sun and the Moon drive the tides of the oceans, and the Sun is the dynamo that generates our weather systems, both on a day to day basis and via orbit and spin variations, over the millennia, of the Earth itself. Scientists happily acknowledge that our understanding of these

Our own Milky Way Galaxy is a spiral galaxy similar to M33, above.

matters is partial and inexact, and present-day doctrines are still shot through with mysteries, aspects of the real world that cannot be explained.

Ancient people had different models of the Universe from ours, with different mysteries. Their understanding was less complex and sophisticated, but no less successful a description of the world as they conceived it; ancient concepts merely had fewer facts to explain.

It is a fallacy of every age to imagine that modern understanding has, somehow, only just broken through the barrier of ignorance—that the latest theory put forward is *the* theory, the one that stops further questioning. Physicists, for instance, thought they were in that position at the beginning of this century—until the great scientist Albert Einstein, with his Theory of Relativity, destroyed their feelings of self-satisfaction.

We may believe we know how cosmological influences affect our planet and our lives, but in fact the areas of our almost total ignorance are great. For example, we have only the vaguest grasp of the ways in which cycles of magnetic perturbation deep within the Sun drive the solar activity seen as flares, prominences and sunspots, which themselves impinge on weather, radio reception and, doubtless, more subtle activities such as animal navigation.

Horoscopes, radio telescopes, and the profound mathematics of the ''Big Bang'' theory are simply three different ways in which human beings try to come to terms with their awe in the face of the immensity and mystery of the Universe. Mysteries, though, should not frighten us, for, as Albert Einstein himself said, ''The most beautiful thing we can experience is the mysterious.''

The light of the world

It is hardly surprising that all races and all peoples have at some time or another worshipped the Sun, our nearest star. What could be more rational than to formalize in a religion the awe anyone must feel in the face of such mighty powers? An entity that gives light and heat with unfailing regularity, makes crops grow, turns night into day, has many of the attributes of a god. All modern science has done has been to suggest mechanisms for the Sun's spectacular power and influence.

Coupled with a knowledge of the links between Sun and Earth has been the ever-present human desire to understand the origins of both bodies. Where did the world and the Sun come from? This is a question—and a mystery—that all peoples, each religion and every thinking person must at some time have confronted.

Almost all the myths and legends about the beginning of the world link its origins with that of human beings. The driving force for this dual creation story is almost inevitably a god figure—often one related in some way to the Sun. Modern models of creation do not necessarily invoke the need for a god to drive the process of creative activity, but present-day concepts are no less awe-inspiring for that simplification. Cosmologists think that the Sun and the Earth arose almost simultaneously around five billion years ago, and that a vast cloud of gas and dust was the origin of the entire solar system. Under gravitational attraction the cloud began to fall in on itself, with more and more material concentrating centrally.

This process eventually generated a rotating mass with a bulge in the middle. The bulge continued to be compressed gravitationally into a denser and denser body until eventually temperature and pressure were high enough to induce thermonuclear fusion—the same process that is at the heart of the hydrogen

The sight of the Sun rising out of the mists of night to give a perfect autumn day encapsulates both the mystery and majesty of this great heavenly body and the Earth's utter dependence on it in every aspect of life.

bomb. A star had been born which began to pour out a stream of heat and light energy that has continued undiminished for more than four thousand million years. The remaining disk of material from the cloud defined the orbital plane of the planets-to-be of the Solar System, and the dust and gas of the disk consolidated itself into the planets.

The might of the Sun's power source has been a mystery ever since people looked into the day-time sky and wondered about the origin of the Sun's apparently inexhaustible heat and light. The ancients supposed that it must be fed by enormous fires, and even by the start of the twentieth century our conception had not advanced very much.

In 1908, Hermann Helmholtz, the American physicist, considered the implications if the Sun were made of a mixture of hydrogen and oxygen and produced its light and heat by the combustion of these two gases. This idea had some credibility at the time, for it was known that the Sun consisted mainly of hydrogen. Helmholz found that, though the Sun's diameter is a staggering 1.3 million km/808,000mls, or more than 100 times the diameter of the Earth, it would run out of fuel in about 3,000 years if it were using it in this way. As the Sun has existed for more than four billion years, this explanation was quite obviously untenable.

Twenty years before Helmholtz's calculations, the British astronomer Sir Norman Lockyer had attempted to solve the mystery of the Sun's enigmatic power source by an even more exotic suggestion. He postulated that all stars derived their power from the constant bombardment of their surface by huge numbers of meteorites; but this was also quite inadequate in practice to maintain the eon's-long energy output of our star.

It was in 1914 that an American astronomer, Henry Russell, and a Dane, Ejnar Hertzsprung, moved the story of a star's power source nearer to our modern understanding. They published a description of the relationship between the brightness and colour of stars as they age—the so-called Hertzsprung-Russell diagram. These findings

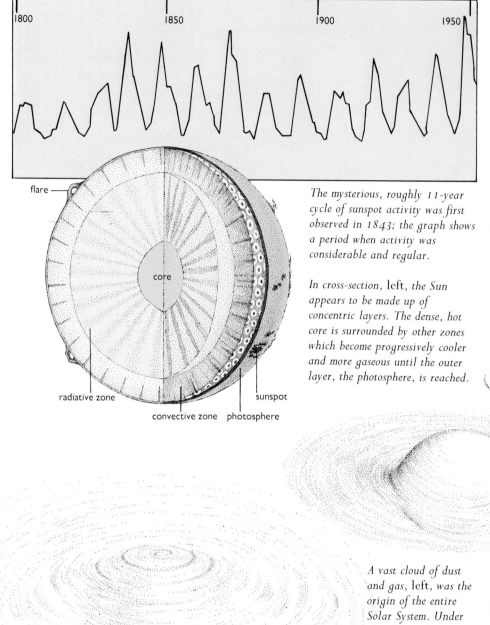

The mysterious, roughly 11-year cycle of sunspot activity was first observed in 1843; the graph shows a period when activity was considerable and regular.

In cross-section, left, the Sun appears to be made up of concentric layers. The dense, hot core is surrounded by other zones which become progressively cooler and more gaseous until the outer layer, the photosphere, is reached.

flare

core

radiative zone

convective zone   photosphere

sunspot

A vast cloud of dust and gas, left, was the origin of the entire Solar System. Under gravitational attraction material concentrated at its centre, creating a rotating mass, above, like two sombrero hats, brim to brim, or two fried eggs.

directed attention to energy-producing mechanisms in stars; Russell at first thought that the energy came from positively charged protons and negative electrons destroying each other. Although this proved to be incorrect, it was nonetheless a move into the appropriate area of possibilities.

The final breakthrough came in the

years immediately before World War II. In 1927, Sir Arthur Eddington, knowing that the main two gases in the Sun were hydrogen and helium, suggested that the energy source was the transmutation of elements. At high enough pressure and temperature, atomic nuclei of hydrogen could merge to make nuclei of helium, and this process of nuclear fusion would

The central bulge became compressed into a denser and denser body until temperature and pressure were high enough to bring about thermonuclear fusion, below. The star that had been created began to emit huge amounts of heat and light. The rest of the material from the cloud condensed to form the planets, right.

The spectacular prominences, clouds of denser and cooler material, that occur in the gaseous outer mantle of the Sun are magnificent indicators of solar activity. This false-colour, extreme ultraviolet image, photographed from the Skylab space station in 1978, was one of the largest ever recorded and erupted for about two hours.

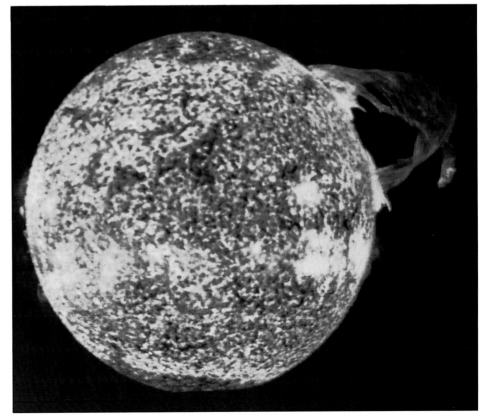

release gigantic amounts of energy. Calculations showed that if the hydrogen was "burned" in this way, it could last for several billion years.

Eddington's fusion theory was refined in 1939 by the American physicist Hans Bethe, who grasped that the small number of carbon nuclei in the Sun could act as a form of self-replenishing

Sun worship

The role of the Sun is paramount in almost all the creation myths of the world. And in cultures as diverse as those of the early Celts in Britain and Gaul and the Aboriginals in Australia, the Sun was worshipped as a god.

The ancient Egyptian creation legends interweave most of these mythic strands. Ra was the sun god, whose name signified the creator and lord of the skies. He was venerated above all other gods and over the centuries had several manifestations, including the hawk, the lion and the sun disk. The human race was created from the tears of the great Ra, the pharaohs claimed to be descended from him, and among his divine offspring were Shu—the Atlas of Egyptian mythology—the earth god Geb and his wife, the sky goddess Nut. This trio comprises the main actors in this typical creation story.

While Geb and his wife were locked in divine intercourse, Shu forced them apart, Nut upward to make the sky, Geb below to create the land. Nut is often represented as an elongated female form, touching the Earth with her toes and fingertips and with her star-spangled belly facing the Earth beneath. In an age when there was no possible way of obtaining realistic information concerning the Earth's origin, these deistic myths provided a coherent and awe-inspiring explanation for that civilization.

Thousands of years later, the Spanish conquerers of Central and South America during the sixteenth century found that sun worship was still an integral part of the religion of the Incas and the Maya. The Incas believed that Inti, the sun god, dived into the sea each evening, to swim under the Earth and re-emerge in the east next day.

This magnificent gold mask from Ecuador, the face ringed with flames, represents the Inca sun god. The priests of the equally Sun-centred Maya religion were skilled at astronomical observation, and at the city of Chichen Itza in Mexico, right, four flights of 365 steps— one for each day of the year—lead to the top of the Castillo pyramid.

The ancient Egyptian Book of the Dead shows one of Ra's priestesses and Hapy, an ape-headed god whose duty was to guard one of the funerary jars, adoring the sun disk, which contains the eye of Ra.

catalyst to assist hydrogen fusion. This forms the basis of the modern view of stellar energy production.

Linked with this view of the Sun's dynamics is a grasp of its internal organization—something that can never be perceived directly. Deep in the Sun's heart is a core of fusing hydrogen under conditions of immense temperature and pressure. Here the nuclei of hydrogen atoms are squeezed together at around $15,000,000°C/27,160,000°F$ to form material 12 times as dense as lead. In these hideously energetic conditions, what amounts to a vast and controlled hydrogen bomb is constantly exploding.

The energy emitted from the core works its way by radiation toward the Sun's surface, to the so-called outer convection zone some $100,000km/62,000mls$ thick. From here outward, most energy is transferred by violent convective stirring motions in the gas. The visible part of the Sun is the photosphere, the shining outer surface, with a temperature of about $6,000°C/10,864°F$. Beyond that is the Sun's tenuous atmosphere, the corona, which is much dimmer and is the origin of the "solar wind" of energetic particles that constantly streams from it.

The seemingly perfectly spherical shape of the Sun and its regularly radiant surface are, however, perturbed by a number of features, many of them rhythmical in character and linked to its magnetic properties. Huge flares and prominences are sometimes emitted from the Sun's surface, and relatively dark, cool sunspots appear, move across the surface, and then disappear.

The number of sunspots can be seen to

rise and fall over the years in a cycle of about 11 years, and it is linked to a measurable 22-year cycle of reversals of the Sun's own magnetic field polarity. Many attempts have been made to show that this cycle has correlates on Earth—with the weather, human behaviour or critical historical events—and almost certainly some meterological effects do follow the same time cycle. This is probably nothing to do with the sunspots directly, but is, more likely, related to the increased solar wind activity, itself tied to the Sun's magnetic cycling.

The cosmic position of the Earth in relation to the Sun and other heavenly bodies has always been a profound mystery. The story of the unfolding realization of something approaching the true position is a fascinating tale of massive insights, long periods of confusion, times of false starts and, eventually, the production of a consensus that gives a clear, if daunting, picture of our cosmological grid reference.

It is easy to imagine how the early idea of a flat, non-moving Earth came about, for to the ordinary casual observer there

is nothing in everyday experience to refute this straightforward and apparently sensible position. Indeed, the Greek philosopher, Thales of Miletus, who lived more than five centuries before the birth of Christ, thought that the world was a flat disk which floated on water. His theory was only seriously challenged when it became obvious that certain astronomical observations could not be compatible with this simplistic view of the world.

In particular, specific bright and easily recognizable stars could be seen at certain times of the year in some parts of the then-known world but never in others. The star Canopus, for example, could be seen at Alexandria in North Africa but never appeared in the night sky at Athens. The only simple explanation of this fact was that the land surface was curved, with the result that, in mainland Greece, Canopus was always below the horizon.

Since the Greeks had a mystical regard for the sphere, which they believed to be the perfect solid shape, it was natural for them to extend the idea of a curved surface to the concept of a spherical Earth—an enormous mental leap for humankind.

The spherical Earth model had the additional virtue of explaining why the shadow of the Earth, which caused the eclipses and solstices, was itself circular. Even 41 centuries ago, the periodic occurrences of eclipses and solstices were being monitored by the Earth's inhabitants. For example, Neolithic people built the remarkable monument of Stonehenge in southern England, on what is now called Salisbury Plain. It probably served not only as some form of temple but as an astronomical computer. Sightings on the Sun and Moon through the complex rings of carefully positioned stones almost certainly enabled the prediction of eclipses.

Even when the idea of a solid, spherical Earth gained credibility, it was still difficult for early thinkers and philosophers to escape from the most tenacious of preconceptions—that humans and the Earth we live on must be at the centre of the Universe. The understandable classic Greek view was that the Sun, Moon and stars must be revolving around a stationary Earth along circular paths: the idea of the heavenly spheres.

The movements of the planets, the Greek *planaomai* or "wanderers", were less easy to explain because their tracks were irregular against the background of "fixed" star positions. But the ingenuity of the Greeks was capable of surmounting this difficulty, and they produced a theory, eventually called the Ptolemaic System, which codified planetary tracks

and involved circles within circles: the so-called epicycles. When appropriately organized, this cosmological set of cogwheels could approximately predict the movements of the planets and stars.

However, some Greeks , the intellectual mavericks of their day, challenged the orthodoxy of an Earth-centered Universe. Aristarchus of Samos was, around 250 BC, the first to suggest that the Earth moved around the Sun along a circular orbital track that took a year to complete. He also concluded that the Sun was larger than the Earth, and that the rotation of the Earth on its own axis caused night and day; for his perspicacity

he was regarded as a deluded eccentric.

It was not until 1530, some 1,800 years later, that this insight was authoritatively restated by the Polish priest, Mikołaj Kopérnik (1473–1543), better known as Nicholas Copernicus. His concept of a Sun-centered Universe immediately aroused the violent opposition of the Church. Martin Luther, the leader of the Protestant Reformation, said of Copernicus's work, "This idiot is trying to overturn the whole art of astronomy". Indeed he was. He had effected an irreparable crack in the edifice of the Ptolemaic System and made its destruction ultimately inevitable.

Andreas Cellarius's Harmonia Macrocosmica, *published in Amsterdam in 1708, depicts the Copernican system of the Universe in an imaginative fashion.*

Circling around a benign Sun, an outsize Earth presents four aspects of night and day, the planets wheel in their orbits, and the 12 signs of the Zodiac appear on the ecliptic—the Sun's apparent yearly path among the stars.

Early astronomers

Nicholas Copernicus's treatise *On the Revolutions of the Celestial Orbs*, published in 1543, laid the foundation for all modern astronomy. However, its author wrongly believed that the planets' orbits were perfectly circular, a notion corrected by the German Johannes Kepler (1571–1642), who found they were elliptical.

The invention of the telescope enabled the Italian astronomer Galileo (1564–1642) to discover, among other things, the craters and mountains on the Moon, Isaac Newton (1643–1727) built the first reflecting telescope, and in 1789 William Herschel (1738–1822), already the discoverer of the planet Uranus, completed his "40ft (12.2m) reflector". With this giant instrument, he swept the heavens and penetrated far beyond our Solar System.

On the basis of such seminal early work has been built today's fantastic successes, such as the Voyager 2 probe to Neptune.

The excitement caused among scholars and astronomers by the ideas of Copernicus is charmingly conveyed in this woodcut, which shows a man breaking through the vault of heaven to investigate its mysteries.

The mighty magnet

For most of us, there is commonsense science and there is mysterious science. Commonsense science has to do with normal expectations, common experience and what we fondly hope is a grasp of the mechanisms that make things happen. We are not surprised when a flame heats a pan of water or when cold water added to hot water makes twice the volume of warm water.

Mysterious science is about those processes which we cannot imagine taking place in our everyday, well-ordered world. At the level of quantum physics—which deals with strange uncertainties on the plane of being where matter and energy occur in infinitesimal, unpredictable amounts—the mystery can safely be ignored. The quantum does not intrude into our workaday affairs.

Magnetism, however, is a different matter. It seems supremely common-place, something to be taken completely for granted. For a few pennies you can buy children's toys that work because of small magnets repelling each other, and

With truly amazing accuracy, migrant birds, such as barnacle geese, return to the same places year after year. They seem to navigate by using the Sun and stars, landmarks, and the Earth's magnetic field, of which humans are only dimly aware. We cannot understand this innate power, but exploit it both for our entertainment, as with these racing pigeons awaiting release, and to carry messages when no other form of communication is possible, as in wartime.

a compass with a magnetic needle in it has been a common and crucial artefact in almost all technological societies. But the toys and the compass, if we stopped to think about them, could terrify us. The capacity of one magnet to make another move, with no physical contact between them, seems little short of magical.

The mysterious powers of electromagnetism, the blend of electrical and magnetic influences that are always intertwined, are of all-pervasive significance. It is electromagnetic attraction, that is, the attraction of a positive charge for a negative one or a north pole for a south, that makes an electric motor or a power station dynamo work. The same attraction, on staggeringly more minute scale, holds every atom together.

Each atom is organized like a miniature Solar System. The "sun" in this system is the atom's nucleus, which contains almost all the atom's mass and is positively charged. Around the nucleus are negatively charged electrons, held in various "orbits" by electromagnetic forces. Without that attraction, which holds the components of the atom together, it could not exist. And without atoms, worlds, suns, and people could not exist.

Our planet has a significant magnetism associated with it, simply demonstrated by the way a compass needle aligns itself with the Earth's magnetic field. In the Northern Hemisphere, the needle points to the north magnetic pole; in the Southern Hemisphere to its southern equivalent. The Earth, in magnetic terms, behaves almost as though it had a gigantic bar magnet stuck through the middle, with a pole at each end of the magnet. Each of these poles is displaced by about 11 degrees from the true geographical poles, which are defined by the axis on which the Earth spins.

Between the two magnetic poles stretch the unseen lines of force of the planet's magnetic field. This force field provides an invisible, patterned, map-like grid in all landscapes for those living things that have the power to perceive its strength and direction. Scientists are only just beginning to realize how

Sir James Clark Ross (1800–62), with his uncle John Ross, in 1831 located the magnetic north pole. In 1839 he led an Antarctic expedition, sailing farther south than anyone else had done, but he did not find the south magnetic pole. The Ross Sea was named for him.

Magnetic north pole   Geographical north pole

Spin axis

The Earth's magnetic field is much like that produced by a simple dipole magnet, right, but it is greatly distorted by the solar wind into a tear-drop shape, opposite.

The dipole lies at an angle of 11° to the Earth's axis of spin, so the true and magnetic poles are not identical. The magnetic poles also wander, following circular paths about 160km/100mls in diameter; at present magnetic north is situated at 79°N, 70°W, northwest of Thule in Greenland, above right.

common such a magnetic sense is among living things. Bacteria, fish, birds and perhaps even people, derive part of their ability to maintain a particular course from their capacity to sense the direction of the Earth's magnetic field. And it is thought that dowsers, people who have the mysterious power to find substances underground using only a rod or forked stick, may succeed because they are sensitive to small electromagnetic variations produced by hidden water or minerals.

Researchers have still only the vaguest idea about the true source of the Earth's magnetism. The fact that it can vary in strength and direction from year to year and from millennium to millennium suggests very strongly that it cannot be due to a rigid magnetized

The magnetosphere

The Earth's magnetic field stretches out into space for many ·times the radius of the planet. Its shape is not the symmetrical, ring-doughnut (toroidal) pattern produced by a simple bar magnet because it is hugely distorted by the solar wind—the never-ceasing stream of charged particles that emanates from the Sun. This compresses the field on the side near the Sun and elongates it enormously on the opposite side. The result is a magnetic field pulled out into a stretched tear-drop shape, with its tail directed away from the Sun.

The field is known as the Earth's magnetosphere, although it is far from spherical, and its boundary is termed the magnetopause. This frontier of the Earth's magnetic influence is only 60,000km/37,300mls out into space on the solar side, but around 400,000km/248,500mls on the side away from the Sun. The great velocity of the charged particles in the solar wind, which may reach 1,000km/621mls per second, produces a type of magnetic shock wave, rather like the bow-wave of a ship, at the sunward boundary of the magnetosphere.

The Earth's magnetic field acts as a snare for some charged particles from the Sun and also for others, known as cosmic rays, from deep interstellar space. The charged particles are trapped in two belts around the planet, one that extends 1,000–5,000km/620–3,100mls, the second 15,000–25,000km/9,300–15,500mls above the Earth's surface. The presence of these zones, called the Van Allen radiation belts after the American scientist who discovered them, was revealed in 1958 by Geiger counters on board the first American satellite, Explorer 1.

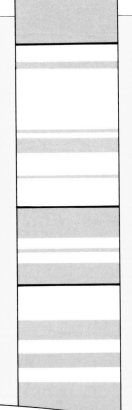

The reversal of the Earth's magnetic field, north for south and vice versa, has occurred throughout time. The record of magnetism in rocks over the past 4.5 million years shows this clearly. The alternating grey and white bands illustrate the many reversals that have taken place over this period.

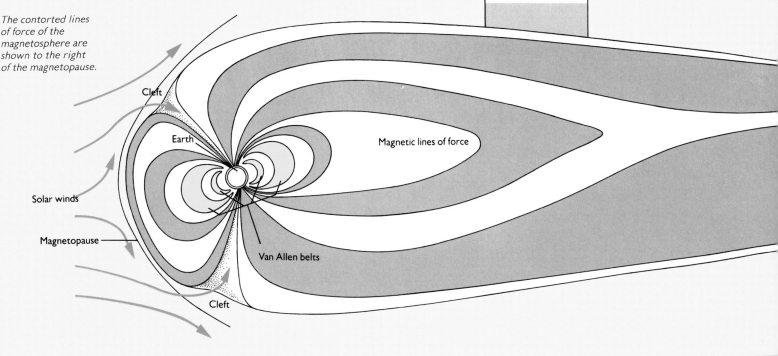

The contorted lines of force of the magnetosphere are shown to the right of the magnetopause.

Cleft

Earth

Magnetic lines of force

Solar winds

Magnetopause

Van Allen belts

Cleft

structure deep in the Earth. The best guesses, and they must remain for the present just that—informed speculations—are that the magnetism is generated in some self-sustaining way by the heat-induced movements of iron-rich metals in or near the Earth's core.

The core of the Earth is approximately 3,600km/2,250mls in diameter and consists of a solid central region and an outer zone thought to be fluid in nature. The iron component of this core metal is a good conductor of electricity, and it is always the case that electric currents in conductors lead to magnetic fields while, when conductors move in a magnetic field, a current is induced in them. This beautiful symmetry of interactions between current, movement and magnetism suggests one way in which the core might be able to set up the planet's magnetic field.

It is called the theory of the self-exciting dynamo. If the conducting, fluid iron of the core is in constant heat-generated circulation and it encounters any slight magnetism, a current will be set up in the moving metal. In its turn, the current will set up a magnetic field that will stimulate even more current.

Such a system demonstrates positive, self-sustaining "feed-back"; and if the circulation of metal is perpetually driven by heat convection, it is possible to envisage the perpetual production of a magnetic field. All that is needed to initiate this process is some original magnetization. As both the nickel and iron of Earth's core can easily be magnetized, it is not difficult to imagine a fundamental magnetism in the solid core setting up such a self-exciting process in the ever-moving fluid core.

Without doubt, the most crucial impact of the Earth's magnetism on humans has been through the use of the magnetic compass for long-distance navigation. A magnetized needle which is able to rotate freely will line up with the Earth's magnetic field and thus, in the Northern Hemisphere, will point to the magnetic north pole.

To be able to ascertain direction in featureless land- or seascapes, when neither Sun nor stars is visible, is a great power and one made use of by all the early seafaring races. The ancient Greeks knew of the strange magnetic properties of lodestone—a strongly magnetic natural mineral consisting mainly of iron compounds—but it was probably the Chinese, who, no later than the first century BC, first constructed a compass with lodestone.

In the end, though, a compass is no better than the lines of magnetic force that direct it. These, unfortunately, are not the same throughout the world. One problem is that caused by the discrepancy at any point on the Earth's surface between magnetic north and true north. This declination varies in a rather complex way and only if the pattern of change is clearly understood can compass readings be turned with certainty into true bearings.

In the United Kingdom, for example, compass north is about 10 degrees west of true north, whereas in the middle of the Mediterranean the declination discrepancy approaches zero. The closer one gets to the magnetic poles the more major are the declination errors—deviations of a compass needle east or west of the true north. This has always been one of the intrinsic problems for polar explorers.

Not only does declination vary with geographical location, the positions of the magnetic poles wander through time; and the strength of the field varies. These complicated changes mean that nautical charts must be updated regularly with respect to the direction of magnetic north, to prevent inaccuracies in navigation accumulating.

Such practical aspects of terrestrial magnetism point up a profound truth about the phenomenon itself. It is by no means the static and fixed grid of magnetic directions that a school atlas would suggest. Rather, the Earth's magnetism is a dynamic entity, no more constant nor unchanging than the climate, the landscape or the positions of the drifting continents. Indeed, considered over very long time intervals, an extraordinary magnetic variation comes to light, which was first noted as evidence of continental drift itself. This is the evidence of paleomagnetism, the magnetism remaining in rocks from times long past.

When molten rock first solidifies, tiny magnetic particles within it align with the existing magnetic field, pointing at either the north or south magnetic pole. Study of rocks which solidified over the past 50 million years shows that at irregular time intervals, typically hundreds of thousands of years apart, the polarity of the Earth's field has inexplicably reversed, the magnetic north pole becoming the magnetic south pole and vice versa. Nothing could show more clearly the dynamism of the Earth's mysterious magnetism than this most startling of changes.

Late 17th-century brassbound lodestone, above.
Chinese geomancer's compass, right.

Development of the compass

The discovery of the magnetic quality of lodestone, or magnetite, and the development of the mariner's compass were the catalysts that prompted navigators to embark on ever longer and more ambitious voyages.

A suspended lodestone will point more or less north–south, with the same end always to the north. Early Chinese compasses incorporating lodestone were used not for navigation but for geomancy, a form of fortune-telling in which the magically turning compass needle was allowed to point at prophesies arranged around it.

Sir Francis Drake, the Elizabethan privateer, was the first man to sail around the world, by way of Cape Horn, the East Indies and the Cape of Good Hope, and live to tell the tale. As this replica of his ship **Golden Hind** shows, his 100-tonne, 60m/100ft long vessel was tiny by comparison with modern ships, but her captain, equipped with the most up-to-date navigational aids, including the compass, had a thirst for adventure and booty that spurred him on and brought him home in 1580, after a three-year voyage.

North was shown by three sets of broken lines, or trigrams, in the innermost circle, and south was, unusually, located at the top of the compass.

Gradually compasses became more sophisticated. Lodestones were used to stroke iron needles, turning them into magnetic compass needles which, allowed to move freely, align themselves with the magnetic north and south poles. In time the needles were carefully pivoted, protected from the wind, and, eventually, suspended in liquid to cut vibration and give a smooth, clear directional response.

This brass-gilt navigator's compendium, made in 1569 by Humphrey Cole, may have belonged to Sir Francis Drake. The slightly oval-shaped instrument is only 65mm/2½in long, but it incorporates a universal equinoctial dial, table of latitudes, tide table and calculator, circumferentor or compass, perpetual calendar and a dial showing the phases of the Moon.

Northern lightshow

The lights of the aurora borealis are among the grandest natural pageants the human eye can perceive. Shimmering arcs and bands of violet and blue play across the dark of the polar night; bright green rays with flashing red tips rush across the sky; and exquisite white draperies with fantastically intricate structures change and merge several times a minute. Meanwhile, patches of pulsating light create a sight whose beauty approaches that of a splendid sunset, but which is in constant motion.

Although its cause is now reasonably well established, the aurora borealis remained a scientific curiosity for hundreds of years, and fascinating mysteries still surround its existence. Unlike the rainbow, whose position appears to change depending on where the observer is, the aurora always comes from definite places in the upper atmosphere, which have the appearance of flamelike arcs or rays. But its curious, unearthly light is not caused by something burning, it is more like the glow given off by electrical discharges in a fluorescent light.

Aurorae—the aurora borealis and the aurora australis—most commonly occur in two bands surrounding the North and South Poles respectively, and usually pass from west to east. The fact that they are nearly perpendicular to the direction of the compass needle gives the clue to a connection with the Earth's magnetic field.

Arctic skies, particularly in northern Canada, Alaska, north Norway and Spitzbergen, are the setting for the most magnificent displays, because they are much darker and clearer than skies over the densely populated regions of Europe. The best time for observing the aurora borealis is in February, when areas of barometric high pressure remain stationary over the north polar regions for weeks on end.

The capricious beauty of an auroral display was marvellously captured by the American poet and writer Bayard Taylor in 1864. ". . . a broad luminous curtain [fell] straight downward until its fringed hem swung apparently but a few yards over our heads. It . . . hung plumb from the zenith, falling, apparently, millions of leagues through the air, its folds gathered together among the stars and its embroidery of flame sweeping the earth and shedding a pale, unearthly radiance over the waste of snow."

During this period, aurorae can be seen on almost every clear night, although they are much less obvious in moonlight. The brightest stars can be seen through auroral light, but brilliant displays give enough light to read by.

An aurora normally appears as a long wavy band or curtain, although sometimes it is merely a diffuse, formless but luminous, mass. If the aurora is almost overhead, viewed edge-on the curtain is seen to be exceedingly thin and deep, on occasion extending upward as much as 650–800km/400–500mls, though some aurorae have a height of as little as 32–48km/20–30mls.

The highest aurorae usually occur in the layers of the atmosphere that are within the Sun's rays, even when it is below the horizon. Due to the curvature of the Earth, those aurorae that appear to be low on the horizon are, nevertheless, high in the sky, but hundreds of miles away.

Individual displays may seem to consist of a random succession of beautiful and delicately tinted forms, but a typical auroral substorm (which usually coincides with a magnetic disturbance) does, in fact, follow a regular pattern through five distinct stages of development. The first indication of an auroral display is normally the appearance of an arc of green light—the quiescent arc—in the northern sky soon after sunset.

This arc is formed by a vertical sheet or curtain of light only a few hundred yards thick, which follows a line of geomagnetic latitude. It can reach for hundreds, even thousands, of miles and usually persists for an hour or so with little change. If the magnetic disturbance dies out, the arc fades away, but if it intensifies the display enters the stage of the active arc.

The lower edge of the arc now sharpens and brightens strikingly, becoming bluish, and moves rapidly southward. At the same time, the form of the arc breaks up into parallel rays or bundles of rays, extending upward to the zenith, which usually drift westward along the arc. If the display intensifies yet again, the third stage begins.

This, the auroral corona, is the most spectacular part, although short-lived. The auroral curtain is now nearly overhead and, looking up into it, a circular crownlike object can be seen, with the rays and flutings along it appearing to converge. Occasionally the corona merges into a bow or fan of light arching across the sky; at other times it throbs rapidly and emits thousands of rays like showers of falling arrows.

After the corona has faded, there is a period of erratic auroral activity, known in the Shetland Islands, to the north of Scotland, as the time of the "Merry Dancers". The display now consists of bands or patches of light, which fade and reappear in a pulsating pattern, sometimes accompanied by flaming, the most breathtaking of all auroral shows.

The fact that aurorae are connected with the magnetic field of the Earth does not, of itself, explain their cause. When the Greek philospher Aristotle observed their brilliant pulsating hues, he concluded that the air was being turned into liquid fire. But it has been believed for many years now that

By studying photographs of aurorae, taken at the same time from two or more observation points linked by telephone, Norwegian Carl Størmer was able conclusively to determine their actual location in the atmosphere.

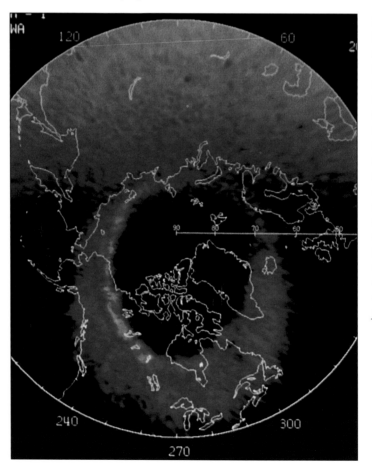

Recorded from space by the Dynamics Explorer 1 satellite, this amazing image-enhanced picture shows a loop of the aurora borealis encircling the north magnetic pole. The zone extends farther south in North America than it does in Europe. So although Stockholm is much farther north than New York, for example, both have an average of 10 auroral sightings a year.

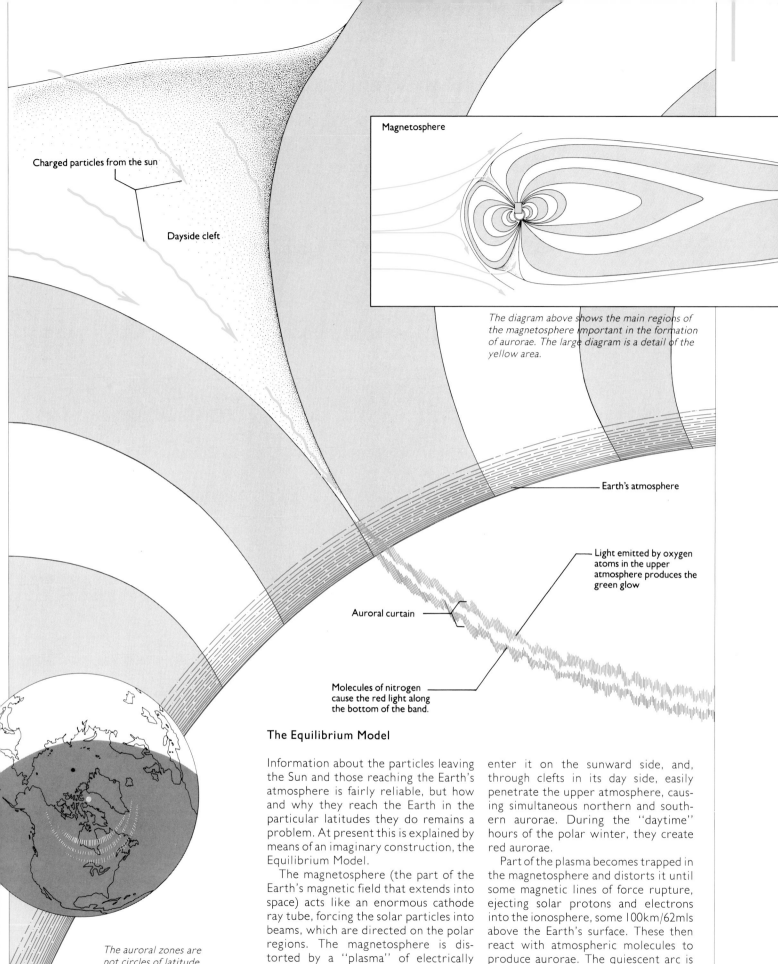

Charged particles from the sun

Dayside cleft

Magnetosphere

The diagram above shows the main regions of the magnetosphere important in the formation of aurorae. The large diagram is a detail of the yellow area.

Earth's atmosphere

Light emitted by oxygen atoms in the upper atmosphere produces the green glow

Auroral curtain

Molecules of nitrogen cause the red light along the bottom of the band.

The auroral zones are not circles of latitude concentric with the geographical poles, but are oval, centered on the magnetic poles.

The Equilibrium Model

Information about the particles leaving the Sun and those reaching the Earth's atmosphere is fairly reliable, but how and why they reach the Earth in the particular latitudes they do remains a problem. At present this is explained by means of an imaginary construction, the Equilibrium Model.

The magnetosphere (the part of the Earth's magnetic field that extends into space) acts like an enormous cathode ray tube, forcing the solar particles into beams, which are directed on the polar regions. The magnetosphere is distorted by a "plasma" of electrically charged atomic particles, electrons, protons, and ions, forming the solar wind.

Most particles are diverted around the outside of the magnetosphere, but some

enter it on the sunward side, and, through clefts in its day side, easily penetrate the upper atmosphere, causing simultaneous northern and southern aurorae. During the "daytime" hours of the polar winter, they create red aurorae.

Part of the plasma becomes trapped in the magnetosphere and distorts it until some magnetic lines of force rupture, ejecting solar protons and electrons into the ionosphere, some 100km/62mls above the Earth's surface. These then react with atmospheric molecules to produce aurorae. The quiescent arc is thought to show where solar electrons are passing down, along the lines of force of the Earth's magnetic field, into its upper atmosphere.

Some Canadian Indians believed aurorae were messengers from the gods; and a "blood red" display over Nuremburg on 5 October 1591 was thought to be a fire in the heavens. Later, the work of astronomers such as Halley and Celsius increased interest in the phenomenon, and in 1838–40 a French scientific expedition went to Norway, where the lower engraving was made.

aurorae are produced by particles emitted by the Sun travelling at speeds so immense that they are able to penetrate deep into the Earth's upper atmosphere, the ionosphere. The invasion of these rapidly moving particles excites the air molecules, which become luminous, giving rise to the auroral light. Different types of aurorae are generated by different types of particle.

Brilliant aurorae, covering the sky and exhibiting the full cycle of stages, are produced from solar flares, which develop from active areas of the Sun's surface. As a result, they vary in intensity with the sunspot cycle, the most frequent and dazzling displays occurring two or three years after periods of maximum sunspot activity.

There are many matters still unsolved by the scientific theory of the aurora. The cause of the innumerable transformations of the delicately coloured shapes and fine structures seen in auroral displays is still only guessed at. And the theory cannot explain an intriguing feature reported by observers: the presence of aurorae between themselves and distant mountains.

The aurorae became of real practical importance in the 1920s, when radio waves were bounced off the ionosphere for the first time in order to extend the range of radio communications. Aurorae were found to absorb some of the radio signals and to cause peculiar reflections of others. Today, however, radio signals are transmitted from satellites in high orbits, so they pass directly through the ionosphere and are much less affected.

Several questions about the effects of the aurora are even now unanswered. For instance, transient electrical fields in the high atmosphere often accompany displays and, in some way, induce electrical currents at the Earth's surface. These currents interfere, in turn, with teleprinter and telephone transmissions over long land lines, or cause instruments used to locate oil or mineral deposits to read incorrectly. Strong currents can even throw circuit breakers in power grids, causing blackouts, as they did in Quebec, Canada, in March 1989.

Even more strangely, many people claim to hear crackling or rustling sounds during violent auroral displays. These noises are not caused by sound waves generated by the aurora, but they may possibly be produced at ground level by as yet undetected electrical and/ or magnetic phenomena that accompany displays. The aurora retains its mystery.

Aurora was the Roman goddess of the dawn, and in 1621 the term aurora borealis was coined by Galileo to describe the breathtaking "northern dawn" or "northern lights". This picture, taken at Fairbanks, Alaska, shows a typical auroral display.

The irresistible force

A baby drops a spoon: it falls to the floor due to gravity. You clip on your skis at the top of the run and move on to the slope: gravitational force starts your downward run. The Moon orbits the Earth, and its circling is driven by the force of gravity. So it is for the planets, comets and asteroids orbiting the Sun; and the same force underpins the majestic circling disk of our Milky Way Galaxy, which is more than 100,000 light years in diameter.

These few examples reveal at once the ubiquity and the omnipotence of gravity. Unlike almost all other forces and influences, it appears to operate with equal power and control on all scales, from the minute to the astronomical. But it took the awesomely great and penetrating intellect of Isaac Newton to grasp its underlying form.

In 1687, quite possibly (as the story goes) after speculating on the significance of a falling apple in his garden, Newton finally published his seminal work, *Philosophiae Naturalis Principia Mathematica* (Mathematical Principles of Natural Philosophy), known as *Principia*. The book covered many topics, but perhaps most importantly it outlined the Law of Universal Gravitation, which stated as a fundamental condition of the Universe that all matter attracts all other matter by means of the gravitational force. The strength of this force depends on the masses of the bodies attracting one another and declines rapidly, in inverse proportion to the square of separation, as the distance between them increases. This elegantly simple and profound statement enabled the prediction of both the fall of an apple and the orbits of the planets.

Gravity is undoubtedly the force that organizes and controls the large-scale behaviour of all components of the Universe. Each large object, be it a planet, star or galaxy, has a gravitational

The Iguaçu Falls, on the border between Brazil and Argentina, provide a spectacular example of the force of gravity. "An ocean pouring into an abyss", in the words of the Swiss botanist Robert Chodat, some 275 individual falls plunge 82m/269ft over the lip of the 4km/2½ml wide rim into the narrow gorge known as the Devil's Throat. The waters then wind their way across the plateau to join the Rio Paraná, 22km/14mls to the south.

field surrounding it. Today, as a result of Einstein's theories, which refined Newton's notion of gravity, the field is often conceived as a "gravity well"—a dimple or depression in the fabric of space-time—with the steepness and depth of the hole determined by the amount of matter in the object. This curving or "warping" of space-time causes any matter or forms of energy, such as light, to bend as they pass through the gravitational fields of massive objects like stars.

Other objects or streams of radiation entering these depressions will slide into them, in other words, will be influenced by the gravitational force. When a spacecraft emerges from the gravity well of the Earth, there are only minor gravitational influences remaining and all objects in the craft become weightless. The weight (mass) of an object is simply the gravitational force operating on it at a particular location.

The scientific advances of the twentieth century have added in many ways to the marvellous edifice that Newton constructed. Again due to Einstein's Theory of Relativity, we now know that mass and energy are equivalent. One can change into the other with a devastating explosion of the atom bomb.

We have also made the imaginative leap to consider objects of such great density that their gravitational well is so deep and steep-sided nothing can escape from it. Such an object is the notorious "black hole"—a cosmological entity whose gravity is so intense it prevents even light from escaping. Perhaps mysterious masses of this type lie at the heart of many galaxies, forming a huge yet invisible core around which each of the myriad suns of the galaxy revolves.

Speculation into the mysteries of gravity has gone further. Just as a moving or vibrating electrical charge emits electromagnetic waves, it has been conjectured that a large, vibrating mass will produce gravity waves—ripplelike disturbances in the gravitational field that move with the speed of light. Many laboratory-based experiments currently in progress seek to reveal the existence of such waves. None of these attempts has yet produced an unambiguously

Sir Isaac Newton (1642–1727)

Mathematician, physicist, astronomer and philosopher, Newton is thought by many to be the most distinguished scientist ever known. He studied at the University of Cambridge and when the university was closed in 1664–66, due to the Plague, he returned to his home at Woolsthorpe in Lincolnshire. Here, at the age of only 23, his most significant discoveries were made.

During this time he formulated his three laws of motion, which constitute the basis of all mechanical science, and evolved the Theory of Universal Gravitation, which shows that the same force governs both falling objects on Earth and the orbits of heavenly bodies. By splitting up sunlight with a glass prism, he found that white light is composed of all the colours of the spectrum; and he commenced work on the calculus, a form of mathematics by which he later explained his scientific ideas.

It was not, however, until 1687 that Newton published his findings in the *Principia*; this was followed by his work *Opticks* in 1704. He became Professor of Mathematics at Cambridge in 1691, President of the Royal Society in 1703, Master of the Royal Mint, and, in 1705, Newton was knighted, in recognition of his services there as well as for his outstanding scientific achievements.

This engraving, after Godfrey Kneller's 1702 portrait, shows Newton at the age of 50.

positive result, but there are great hopes that a recently monitored supernova (a cataclysmic stellar explosion) might emit gravitational waves that will be detectable.

Our ability to probe the mysterious has spawned attempts in the last decade to produce "Theories of Everything", which try to find links between all the forces of nature, including gravity. Present understanding suggests that at ordinary energy levels and conditions four forces operate. These are gravity, the electromagnetic force and the so-called strong and weak nuclear forces, which control the structures of atomic nuclei. These latter two operate over only tiny distances, whereas gravity and electromagnetism have effects at infinite distances.

Gravity is by far the weakest force of the four; the others are all at least 10^{35},

that is, 100,000,000,000,000,000,000,000,000,000,000,000, times stronger. Yet despite gravity's puny effectiveness, it dominates on the astronomical scale, since it can operate at all distances. Electromagnetic forces do not control cosmological movements because the Universe is electromagnetically neutral: every positive charge is counteracted by a corresponding negative one.

Some scientists have argued that at higher and higher energy levels, these four very disparate forces of nature could merge into a single force. Energies of the necessary magnitude to bring this about have, perhaps, only ever existed early on in the "Big Bang", the explosion that started the Universe. If this is so, a mysterious and majestic unfolding of nature occurred in the first moments of that event as the forces were revealed and matter came into being.

The four forces

Gravity
Of the four known "forces of nature", gravity is the one that dominates in controlling the macrostructure of the Universe, underpinning, for instance, the orbits of planets around a star. It does this, despite being the weakest of the four forces, because it operates over infinite distances.

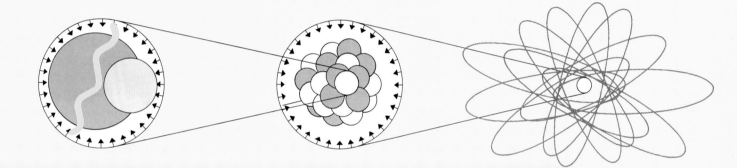

Weak nuclear force
Much stronger than gravity, but working only over miniscule subatomic distances, the weak nuclear force controls the disintegration of some fundamental particles. An example is the break-up of a neutron to give a proton, an electron and the enigmatic, massless and chargeless neutrino.

Strong nuclear force
This force holds together the nuclei of atoms, which are made up of neutrally charged neutrons and positively charged protons. Without it, the like charges on the protons of a nucleus would repel each other, and this would make nuclei—and matter as we know it—impossible.

Electromagnetic force
This is the force of charge and magnetic polarity, in which opposite charges, positive and negative, attract one another. It can operate over infinite distances, but it also stabilizes, for example, the "orbits" of negative electrons around the positive nucleus of any atom.

US astronaut Bruce McCandless floats freely above the Earth in his manned maneuvring unit during the 10th space shuttle flight in February 1984. The MMU, propelled by small, hand-operated nitrogen thrusters, enables the astronaut to move at will instead of just drifting in space, so overcoming the effects of weightlessness, or zero gravity.

Rhythms without end

In 55 BC, Julius Caesar was almost brought to grief by an earthly mystery that must have seemed like a peculiarly effective piece of magic summoned up by his enemies. At the time, he and his expeditionary army were attacking the southern coast of Britain, and his well-trained and disciplined mariners appear to have been fooled by the huge tides that are experienced in the English Channel.

Early in the campaign, these sailors, who were accustomed to the almost imperceptible tides of much of the Mediterranean Sea, did not haul their boats far enough up the beach. Consequently, a very high Channel tide, coupled with and driven by a storm, caused great damage and loss of ships in the Roman fleet. The campaign survived this tide-wrought catastrophe, but Caesar was, doubtless, left pondering what forces could, within hours, push the sea such a great distance up the shore.

Nowadays we believe we know what those forces are, where they originate and why tides vary so greatly in different parts of the world. But there are still aspects of the tidal phenomenon that have the power to perplex and baffle. How, for instance, do sea creatures become trained to the rhythms of the ocean tides? Is it just coincidence that a woman's menstrual cycle has the same periodicity as one of the major tide cycles? Are there atmospheric tides in addition to the tides in the seas? Even to begin to grapple with such problems and possibilities, we must go back to the cosmic foundations of Earthly tides.

Early peoples knew that the tides in the seas were linked in some way to the positions and phases of the Moon, but the explanation of how they were linked had to await Isaac Newton's comprehension of the existence of gravity and its effect on all matter. With the publication of his theory in 1687 came the

A full Moon, hanging over a silver sea, seems the epitome of tranquillity, but the calmness of the scene belies constant activity. For it is the mysterious power of the Moon that is largely responsible for the endless change and movement of the tides in the Earth's oceans and, perhaps, for similar tides in the atmosphere that surrounds it.

realization that gravitational forces operate between all bodies, and that their strength depends on the size of those bodies and their proximity. So it became clear that the tidal movements of the oceans on Earth had a gravitational cause, and that the attraction pulling the waters into high tides and intervening low tides was the result of a complex choreography of Earth, Moon and Sun in the infinity of space.

Both the Moon and the Sun exert a gravitational attraction on the oceans. Although the Sun has a mass 27 million times greater than the Moon's, it is the Earth's tiny natural satellite which has the major effect on the tides: some 70 percent to the Sun's 30 percent. The reason for this is the difference in the distances of the Sun and the Moon from the Earth—the Sun is 150 million km/ 93,226,600mls away, while the Moon is only 380,000km/236,200mls distant on average.

The Moon, then, is the major genera-tor of tides, which it causes by drawing up the waters beneath it into a bulge. As the Earth spins under the Moon, this bulge is experienced as a high tide. In fact, on most coastlines around the planet, an observer sitting at the top of the beach would experience two high tides and two low tides each 24 hours. Superficially, this is an unexpected situa-tion, since it might be imagined that

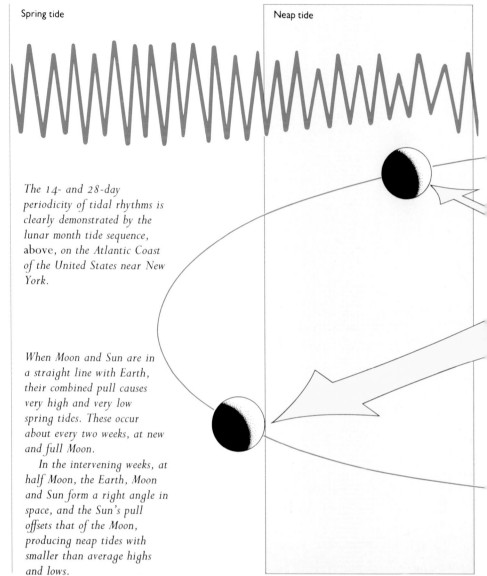

Spring tide   Neap tide

The 14- and 28-day periodicity of tidal rhythms is clearly demonstrated by the lunar month tide sequence, above, on the Atlantic Coast of the United States near New York.

When Moon and Sun are in a straight line with Earth, their combined pull causes very high and very low spring tides. These occur about every two weeks, at new and full Moon.

In the intervening weeks, at half Moon, the Earth, Moon and Sun form a right angle in space, and the Sun's pull offsets that of the Moon, producing neap tides with smaller than average highs and lows.

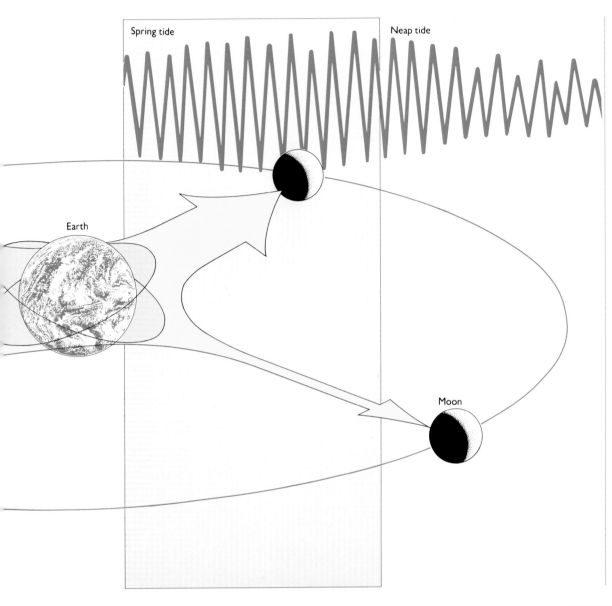

Sun

Spring tide

Neap tide

Earth

Moon

The Moon shines only
by reflecting the light
of the Sun, and the
phases of the Moon
seen by an observer on
Earth are the
changing shapes of its
sunlit face. When the
Moon lies between
Earth and Sun, its
dark side faces the
Earth, and it cannot
be seen.

We see a full
Moon when the Earth
lies between the Sun
and Moon. The
waxing Moon passes
from new to crescent
to half, and then to
full. As it wanes, we
see the gibbous and
waning Moon, its last
quarter and, finally,
the sliver of the old.

with the Moon orbiting the Earth once every 28 days, and the Earth spinning around once in 24 hours, there should be only one high tide approximately every 24 hours.

This apparent puzzle is solved by more detailed consideration of the orbiting dance of the Earth and Moon. The true state of affairs is that the Moon does not, in fact, revolve around our planet; rather, both bodies revolve around a common centre of gravity located on a straight line between their centres. This gravitational centre of the Earth–Moon axis lies inside the Earth, well over on the side on which the Moon is to be found at any one time.

This means that although there is a high tide bulge of water under the Moon, there is also a second, centrifugal bulge on the other side of the planet. It is these two enormous humps of water that cause the typical two high tides a day at most places in the world.

Once the combined effects of Moon and Sun on the Earth's oceans are taken into account, the reasons for other rhythms and patterns in the tides begin to emerge. One of these is the correlation that so impressed the soothsayers, mystics and astrologers of the past—the clear and unmistakable link between the phases of the Moon and the changing heights of the tides over an extended period, for it has oscillations of both 14 and 28 days within it. Perhaps more clearly than any other natural occurrence, this connection shows how cosmic processes may have had direct effects on early inhabitants of the Earth, since it doubtless fueled many mystical speculations about the influence of other heavenly bodies, such as the stars, on human life.

It is the relative positions of Earth, Moon and Sun that produce the phases of the Moon and also generate the pattern of spring and neap tides. Spring tides, with exceptionally high high tides and low low tides, occur at the times of the new and full Moons, a spacing of about two weeks. The fluctuation of neap tides, both up and down, is far less. Neaps occur between the springs, and the total cycle takes roughly 28 days.

This mysterious linking of tide and Moon shape is, yet again, due to gravity. At the spring tide configurations, Earth, Moon and Sun are in almost a straight line, and the gravitational attractions of our two heavenly companions are, therefore, pulling along the same line. Consequently, they produce two greater than average high tide bulges, and the spring tides are formed. Half way between these two tidal peaks, the Sun and Moon are at right angles to one another in relation to the Earth. This means that their gravitational pulls partly cancel each other out, and the lesser neap tides result.

Real tides on real shorelines, and in the middle of actual oceans, have more complex periodicities and irregularities than the simple models above would suggest. This must be so to explain fantastic tidal differences of the type that so confounded Julius Caesar. Other subtle factors, such as friction between the ocean bottom and moving water masses, and the physical blocking of

Tidal bores

Tides are influenced not only by gravity but by the effect of the Earth's spin, the continents and mid-ocean ridges, and the frictional drag of water on the seabed. These combine to produce complex systems in the great oceans, each with its own basin, in which the tides move counterclockwise around a central node once in 12 hours.

Places having high tide at the same time are joined by lines, each of which indicates an hour later than the line behind it. Tides vary in height from nil at the nodal point to more than 12m/40ft in, for example, the English Channel.

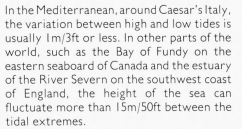

At Lower Parting, some 29km/18mls up river, the Severn bore is around 84m/275ft wide and reaches 16–21kmh/10–13mph. Biggest bores occur around the spring and autumn equinoxes.

In the Mediterranean, around Caesar's Italy, the variation between high and low tides is usually 1m/3ft or less. In other parts of the world, such as the Bay of Fundy on the eastern seaboard of Canada and the estuary of the River Severn on the southwest coast of England, the height of the sea can fluctuate more than 15m/50ft between the tidal extremes.

Both these locations are tapering coast-lines facing a very large ocean, the North Atlantic. In such situations, the funnel shape of the comparatively shallow inlet constricts the huge volume of water moving toward the land with the incoming tide, forcing it up; so the height of the tide inevitably becomes greater.

Where the tapering shape is extreme, a tidal bore results. When the waters racing up the river estuary meet the seaward flow of the river itself, a distinct, steep-fronted wave of water forms, which forces its way far inland, gradually losing its momentum and dying away.

The largest tidal bore in the world is the "Silver Dragon". Once a month, this surge of water, more than 6m/20ft high and 8km/5m wide, roars up the Quiantang River from the China Sea, a cause of potential devastation to the surrounding country-side. Recently, a team of surfers from Britain attempted, unsuccessfully, to ride this mammoth wall of water.

At the top of the Bay of Fundy, an inlet some 270km/170mls long and 50–80km/30–50mls wide, another huge bore invades the small Petitcodiac River and also causes the astonishing sight of falls on the St John River where the water flows upward. Even on such a great waterway as the Amazon, the irresistible power of a tidal bore, up to 4.6m/15ft high, overrides the flow of the river once a month with the spring tide, tearing inland at a speed of 16–24km/10–15mls an hour. And on the Hooghly River, which flows into the Bay of Bengal, there is also a significant bore.

In England, the most impressive bore is that on the Severn River, which occurs with the high tides on 130 days of the year. It reaches its maximum height of some 3m/10ft at the time of the spring and autumn equinoxes, when it penetrates more than 32km/21mls inland, reaching beyond the city of Gloucester. Here, too, there are hardy and intrepid surfers who thrive on the excitement of trying to ride the wave.

water movement by continental land masses, make for changes in tidal type from one sea or ocean to another. For these and other reasons, the relatively small, land-locked Mediterranean Sea has tiny tides by comparison with those around the coasts of the large oceans of the world.

The tides are probably the most potent of all ecological factors for the small creatures and plants that live along shorelines. A limpet, barnacle or frond of seaweed attached at the mid-zone of a tidal shore encounters dramatically varying conditions during a 24-hour period. Twice it will be completely covered in seawater for a variable period; twice it will emerge into the air, perhaps into highly desiccating sunny and windy conditions.

These shore organisms must, then, be extraordinarily adaptable, since they have to change from a marine to a terrestrial lifestyle every few hours. To achieve this adaptability, they have developed a multiplicity of specializations. Many "shut up shop" at low tide, becoming inactive and compact to conserve water inside a sealed shell, carapace or mucus covering. Some have built-in biological clocks, in tune not with the Sun but with the tides, which enable them to attain a protective state just as the falling tide exposes them to the air.

The tides clearly influence these marine animals, but do they have any direct impact on human existence other than the obvious navigational and environmental effects? At first it would seem that the similarity between the lunar and tidal cycle of around 28 days and the 28-day cycle of most women's menstrual pattern is too close to be mere coincidence. Some anthropologists have suggested, for instance, that our species had a long period when we spent a great deal of time in coastal waters, gathering food such as shellfish. They have gone on to conclude that the female reproductive cycle became trained to the tidal cycle because of this ecological linkage.

Most scientists, however, now discount these theories. Human societies have been successful from the earliest

The Californian grunion, Leuresthes tenuis, is attuned to the Moon in a remarkable way. From March to September, these 15-cm/6-in fish swim ashore to spawn at either the full or crescent Moon: spring or neap tide. The largest spawning takes place with the spring tide.

As the high water mark is reached, the grunion start to "run" *and are carried on to the beach by the waves. The pair intertwines, and the male fertilizes the female's eggs as she deposits them in a scrape about 5cm/2in deep scooped out by her body; they then return to sea. Two weeks later, at the next spring tide, the sand-covered eggs hatch and the larval grunion are swept out by the waves.*

The immense tides in the Bay of Fundy scour the coastal cliffs, undercutting them and leaving huge isolated stacks or "flower pots". In their turn, these will crumble, eroded by the waves.

times at inland as well as coastal locations. More tellingly, when the estrous cycles of our evolutionary cousins—the apes and monkeys—are monitored, a wide range of cycle times is found. Rather prosaically, it appears that the 28-day female cycle is only by chance similar to that of the tides.

Nevertheless, other researchers have speculated that the tides do, indeed, impinge on human life in a fascinating and undreamed-of manner. They propose that the Earth's climate may be affected not by the oceanic tides but by comparable, but far more elusive, atmospheric tides. These are small patterns of changing pressure caused by the gravitational pulls of the Moon and Sun on the gases of the Earth's atmosphere.

There is an 18.6-year cycle in Sun–Moon–Earth positions, which enables the straight alignments that cause

eclipses to be predicted. This same type of alignment might also explain the more or less 20-year cycles that seem to be detectable in some climatic data. Rainfall patterns in the American Midwest, for example, show some evidence of such oscillation. If these suggested links are actual, there is every reason to believe that the proximity of the Moon can influence our lives in ways that are only just beginning to be understood.

Wave-driven erosion of a different type is shown on this Dorset beach. Huge tides in the English Channel sweep the shingle diagonally upward and the backwash rolls it straight back down, creating a saw-tooth pattern of mounds and hollows.

Invaders
from space

In the 1950s, at the height of the Cold War between Russia and the West, a spate of sci-fi movies was unleashed on cinema audiences in America and Europe. These movies, featuring alien invasion, gave expression to deep-seated human fears. The invaders came in strange and demonic forms, in innumerable disguises and always with malign intent. And the threats came not from behind the Iron Curtain, but from deep interstellar space.

None of these space raiders has made a documented appearance, but every day, almost unremarked despite our fascination with the subject, real invaders from space waft or hurtle down on to our planet. They, and the remains they leave, are the source of many interlocking scientific mysteries.

The invaders are meteorites—the non-planetary fragments of solid matter that litter the space of the Solar System and finally fall into the gravitational field of Earth. They rocket down through our atmosphere at speeds of up to 70km/44mls a second. At such speeds, air in front of the invaders from space, compressed by shock waves, becomes incandescent. The superheated air heats the outer layers of the meteorites until they, in turn, begin to glow and eventually to melt. The burning gas and molten material thrown out during this process cause the fireball and blazing streak of a meteor across the day or night sky.

The final fate of these fiery intruders depends on their size, composition and trajectory; and their differing fates lead to a confusing multiplicity of names to describe the objects and phenomena. A meteor is what is commonly called a shooting star—a streak of glowing light across the sky—while the object that causes a glowing fireball is termed a bolide, or meteoroid. Many small bolides, which enter the atmosphere at steep angles and high velocities, or are

About 22,000 years ago an immense extra-terrestrial missile moving at some 48,000kmh/30,000mph came into cataclysmic collision with the Earth. It left the huge gaping hole in the Arizona desert that is Meteor Crater. The blast on impact would have been the equivalent of 500,000 tons of high explosive.

Huge craters on the dark side of the Moon bear witness to an awesomely chaotic time millions of years ago, when massive meteorite bombardment scarred its surface.

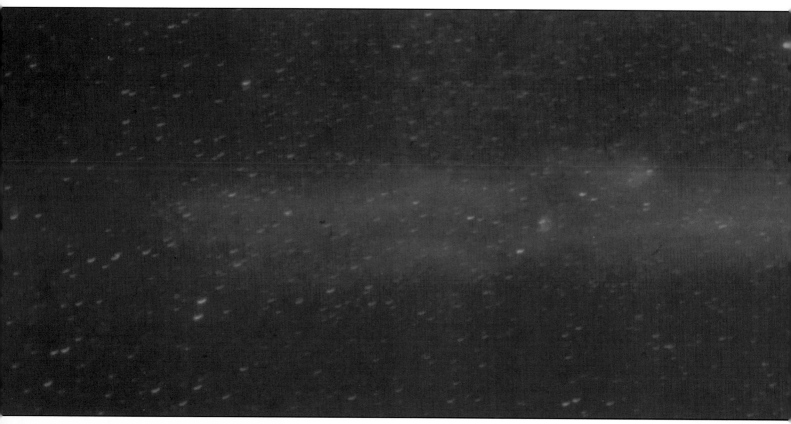

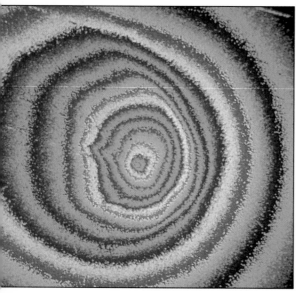

The image-processed optical photograph of the head of Halley's comet, above, is derived from four exposures made in May 1910 at Helwan in Egypt. The colour contours, from green to blue, show the increase in brightness of the comet's nucleus.

Millions of miles of cloud and gas stream out in a tail from Halley's Comet, above right. A curving Type II tail is also visible.

made of easily fragmented material, melt, vaporize and break up during their descent and never reach the Earth.

Only those bolides that do not completly burn up in the atmosphere and actually reach the ground can strictly be called meteorites. Some were huge to start with, so even after depletion by melting enough material is left to impact on the Earth. Others, particularly metallic meteorites, are extremely tough; or they enter the atmosphere on a glancing, oblique trajectory, which means that they lose speed gradually and do not overheat. This "surface skimming" is the same technique used by a returning space shuttle to get back to Earth safely.

Meteorites and meteoroids of all sizes continuously bombard the Earth. It has been estimated that some 20,321 tonnes/20,000 tons of such material gets through our atmosphere every year; that is, more than 50 tons a day! The great bulk of this interplanetary debris is minute, consisting of objects the size of dust particles or sand grains; only a tiny proportion of the total comprises objects of considerable dimensions. The

larger the object, the more unlikely it is that it will produce a meteor.

Unlikely, though, does not mean impossible. Around 50,000 years ago in northern Arizona, long before the feet of humans had trodden the ground of the Americas, one of the rare giant earthfalls occurred. A massive fragment of nickel and iron, perhaps 41m/135ft in diameter and weighing some 305,000 tonnes/300,000 tons, hurtled earthward at about 19km/12mls a second. It must have produced a stupendous fireball, which struck the ground in Cañon Diabolo.

At the impact point, millions of tons of rock and soil were forced into the sky, a gigantically powerful shock wave flattened all vegetation for many miles in all directions, and fires must have destroyed everything flammable for huge distances from the site of the explosion. The meteorite itself disintegrated, the fragments melting and partly turning to vapour.

The result of the impact that we see at Cañon Diabolo is a crater nearly 1.6km/1ml in diameter called Meteor, or Barringer, Crater. The crater was

Meteorite strikes

In the lifetime of the Earth, Meteor Crater in northern Arizona is a relatively recent event. Geological surveys have revealed much older craters that have been greatly changed and eroded over the intervening ages and so are less easy to recognize. One of these is Duolun Crater, only now discovered, near the Luan River on the border of Hei Province and Inner Mongolia, almost due north of Beijing in China. Geologists believe that the meteorite that formed this huge crater, 70km/43mls across, fell some 136 million years ago, as the Jurassic was giving away to the Cretaceous period.

Evidence of an even earlier meteorite strike is the vast 75km/47ml wide Manicouagan Crater in Quebec, Canada, which was created some 210 million years ago. It is calculated that an impact of the scale needed to produce so tremendous a cavity would have melted as much as 1,000cu km/240cu mls of target rock and released as much energy in this one spot as all the earthquakes in the world over 1,000 years.

Events such as those at Manicougan and Duolun must have convulsed the climate of the world. Dust and debris from these explosions must have produced the equivalent of a nuclear winter, with the Sun blotted out and the consequent dying back of vegetation. It is even possible that meteorite explosions contributed to the ulimate demise of the dinosaurs: when the plants died off, these great creatures starved and were no longer able to breed efficiently.

It is also possible that such meteorite falls gave rise to the strange fused objects called tektites, which are found around the world, but predominantly in four large fields in Australasia, North America, Czechoslovakia and the Ivory Coast of West Africa. Tektites, usually weighing only a few grams, resemble volcanic glass in some ways, but most scientists believe that they were once drops of molten earth thrown out by colossal meteor impacts. This theory seems to be borne out by the surface appearance of many tektites, which show clear signs of melting, caused by heating by friction with the atmosphere.

discovered by Europeans in 1891, although it must have been known to the North American Indian tribes of the region before that date. When it was formed, it must have been some 230m/750ft deep, but soil erosion has since reduced this to around 177m/580ft. Remnants of the iron-nickel of the original bolide have been found beneath the crater and within a 10-km/6-ml radius of it.

Where do these objects with the power of thermonuclear weapons come from? Most answers to this mystery seem to suggest that they are fascinating archeological debris from the early stages of the formation of the Solar System, since most meteorites contain mineral crystals which scientific dating tells us must be 4.6 billion years old. At this time, grains of interstellar dust in the cloud, or nebula, that was to form the Solar System gradually coalesced into small bodies known as planetismals.

Early in the evolution of the Solar System, these were in a constant state of collision, accretion and fragmentation. In the larger aggregations, which heated up, there was a difference in the density

As much energy was released when Manicouagan Crater was formed as by all the earthquakes in the world over a period of 1,000 years.

of their structure, and rocky silicate materials floated to the surface while denser iron and nickel-rich materials sank to the core. When such partitioned planetismals were broken up by collisions, some fragments were metal-rich, others rock-rich.

Eventually the planets forming in the Solar System gathered up most of the planetismals by gravitational attraction. Those that remained after this "clean-up" were mainly concentrated in a belt between the orbits of Mars and Jupiter known as the asteroid belt; the asteroids are planetismal debris.

Occasionally the immense gravitational pull of Jupiter perturbs the orbit of an asteroid, nudging it into an eccentric orbit that may cross that of the Earth. Once this happens, asteroid debris has the possibility of becoming meteorites. These may be rocky, metallic or mixed in character, depending upon which type of compartmentalized planetismal they derive from. Meteorites bombarded planets and moons far more commonly in the first half billion years of the existence of the Solar System, when they were more numerous, than they do at the present time.

The only other likely source of meteoritic material are comets. Comets are thought to be "dirty snowballs": compact masses of ice, dust and small rocky fragments. They travel on highly elliptical orbits which take them close to the Sun and then out into the far reaches of the Solar System—sometimes even farther out than the orbit of Pluto, our most distant planet. Some astronomers believe that new comets are projected into the inner parts of the System from a huge collection of ice-rich masses—the Oort Cloud—which is surmised to exist beyond the planets.

Comets are invisible until their core, or nucleus, begins to be affected by solar radiation on the inward leg of the comet's orbit toward the Sun. This causes the comet to heat up, and a cloud of dust and gas, called a coma, is thrown out from the nucleus. As it comes yet nearer to the Sun, a comet forms a tail which, blown by the solar wind, always points away from the Sun: behind the comet on its inward trajectory, in front of it on its outward track. Type I tails are long and straight. Type II tails are shorter and curved and are made up of fine particles of matter forced out of the coma by radiation pressure.

Periodic comets have a wide range of periodicities, or orbit times. Short period comets appear every 20 years or less, medium period ones every 20–60 years, and long period comets at intervals of more than 60 years. Halley's Comet is the most famous example of a long period comet.

Short period comets, at their most distant from the Sun get no farther than the orbit of Jupiter; long period examples journey out far beyond Neptune.

With each orbit, some material from the comet's nucleus is boiled off by the Sun; eventually all that is left is a cloud of orbiting rocky fragments. When these cross the Earth's orbit, they may be the cause of meteor showers.

It seems at least plausible that whole cometary nuclei may occasionally collide with Earth. The Tunguska event—a huge explosion in Siberia in 1908 that flattened trees over a wide area but left no crater—might have been caused by an icy nucleus exploding in the atmosphere above the spot and leaving little residue. The event, like so many other aspects of meteoritic space invasions, is a mystery and one whose cause will probably always elude us.

A regular visitor from space

The great Florentine painter Giotto saw a comet in 1301. That comet, later to be known as Halley's, became the Star of Bethlehem in his fresco The Adoration of the Magi in the Arena Chapel, Padua, which was completed three years later.

St Matthew's Gospel, the only one to mention the "star in the east", was written around AD66, when Halley's Comet was also visible. He may have included the image as a record of his personal experience.

Halley's Comet is the best known of all the periodic comets, largely because it is exceptionally bright and has a long, conspicuous tail. In addition, it appears every 75–76 years—at least once in the lifetime of most people. It was last visible in February 1986, and before that in 1910.

The Englishman Edmond Halley (1656–1742) was the first to recognize that the comet he saw in 1682 was the same one which had been seen in 1531 and 1607. Realizing that this comet had a regular orbital period, he predicted it would be visible once more in 1758–59. He was proved right but, sadly, did not live to see again the comet that now bears his name.

Throughout recorded history, the appearance of Halley's Comet has been noted, and it has often been thought to presage great events. The earliest reference

Hundreds of thousands of meteors shower the Earth's atmosphere every day. Some—shooting stars—are burned up. Others strike the Earth at random and with immense force. Although only 13 large craters have been proved to be the result of meteorite impact, hundreds of sites are known, particularly in North America, Europe and Australia.

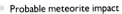 Probable meteorite impact

Definite meteorite impact

is that of the Chinese in 240BC, and it was seen in 12BC and AD66—not as is sometimes thought at the time of Christ's birth.

Belief in the comet's appearance when Christ was born was greatly reinforced by its inclusion in the painting of the Nativity by Giotto di Bodoni (1267–1337). Giotto saw Halley's Comet in 1301, when it had an especially long and fiery tail.

This first realistic impression was nicely commemorated in the naming of the 1986 European space probe. The Giotto craft penetrated the coma of Halley's Comet, sending back detailed pictures, although it did not survive the encounter intact.

The first depiction of the comet was in the famous Bayeux Tapestry celebrating William of Normandy's conquest of England. The Normans regarded the comet, which was visible in 1066, as an auspicious omen, but King Harold and the English are shown trembling at the sight.

The Bayeux Tapestry, right, actually a long strip of embroidery, depicts Halley's Comet as a shuttlecock, stylized but still recognizable.

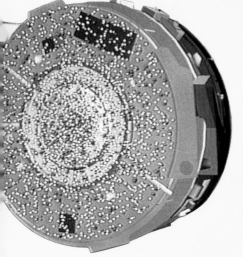

The likely damage suffered by the Giotto probe is shown in this computer simulation of the appearance of its metal shield after high velocity impact with particles of dust.

Cycles of ice

Fears about the "Greenhouse Effect", caused by carbon dioxide produced by burning fossil fuels, are among the deepest anxieties of the developed world. Indeed, most schoolchildren can talk about global climatic change and the factors influencing it with knowledge and assurance. As we ponder the possibility that we might wreak irreversible changes in the weather systems of our planetary home, our perspective has become worldwide and all-embracing. This new viewpoint sets research into climate changes of the past in stark and applied relief: we look to such changes to give us clues to a greater understanding of present ones.

Since the human species emerged, during the last million years, the most profound and challenging climatic mysteries have been the switchback movements in and out of ice ages. What are they? What causes them? Why do they begin, and what makes them end? It is clear that if we can understand the mechanisms underlying these mysteries, our strategic grasp of Earth's climate will be a robust one.

Most of the past three million years—the Pleistocene period—have seen a remarkably rhythmic series of ice ages, or glacials, interspersed with warmer, wetter periods known as interglacials. Over this time, most glacials have lasted around 100,000 years; the interglacials have on average been shorter, lasting 10,000–20,000 years. Today we find ourselves in a relatively warm interglacial, following an ice age that ended just over 10,000 years ago.

If, in imagination, we were to travel back 20,000 years into the last ice age, the Northern Hemisphere in particular would seem totally different from today. The Arctic Ocean and much of the North American and northern Eurasian land masses were buried under ice. In the New World, ice sheets had spread

Evidence of past ice ages is all around us. The largest remaining icefield outside the polar regions is the Columbia Icefield, below, *in Canada's Rocky Mountains. When such great glaciers grind their way down valleys, their erosive force cuts the rocky sides of the valleys, widening and deepening them and slowly changing them to the classic U-shaped glacial troughs, like that of the valley of the River Coe—Glen Coe—in Scotland,* right.

out from their polar strongholds to cover the eastern regions of Canada, Alaska, much of the American northwest, the Midwest and New England.

On the other side of the Atlantic, from Scotland and Scandinavia, ice masses moved south over most of Britain and northern continental Europe. Small ice caps reached out from the Alps and the Pyrenees into the surrounding terrain. And fringing all these consolidated ice sheets, which extended from the mountains in glaciers of huge size, were broad zones of frigid tundra. These massive alterations in the landscape were, in turn, the direct cause of changes in the living creatures to be found: long-haired mammoths, adapted to a polar environment, foraged in lands that now have a Mediterranean climate.

What influences could make our planet's climate oscillate between this frozen world and today's weather? Scientists now think that these mysterious ice-age rhythms are brought about by slight, slow changes in the orientation of the Earth as it orbits the Sun— the ultimate generator of all earthly weather. Crucially, in light of our present concerns over the Greenhouse Effect, these astronomical factors seem also to impinge on a feedback mechanism for planetary temperature control which involves carbon dioxide, the chief greenhouse gas.

Our knowledge of the important astronomical rhythms that affect our climate is based on studies from the late 1860s published by James Croll (1821–77). Croll was a self-taught Scot, a onetime insurance salesman and janitor, who became a Fellow of the Royal Society, the foremost British scientific organization. His studies were updated and cast in essentially modern form by the Yugoslav astronomer Milutin Milankovitch (1879–1958), in the decade before World War II.

The Milankovitch Model, as it is called, suggests three rhythmic alterations in the Earth's celestial activity. These recur at different intervals and could lead to cycles of relative cooling in high latitudes on the Earth. These are, first, a wobble in the spinning axis of the

Glacial erosion

Most obvious relics of past glacials are those of the Pleistocene period, which shaped many landscapes, particularly in the Northern Hemisphere.

The 4m/13ft high cavern at the head of the Franz Josef glacier in New Zealand, below, shows how erosion, a "glacial mill", can gouge out and polish the rock. Then the ice and gravel scour the rocky sides of the valley leaving a permanent record of their passing, opposite.

The debris dumped by bygone glaciers ranges from immense boulders to moraines and drumlins, top. These streamlined hummocks, formed when glaciers shed stones and sand as they spread out on level ground, can be as much as 90m/300ft high and 1.6km/1ml long.

Earth, like that of a spinning top, with a period of around 20,000 years and, second, a cycle of changing tilt of the axis of spin with a 41,000-year period. Finally, there is a much longer, complex oscillation, with periods of 100,000 and 400,000 years, caused by the changing shape of Earth's orbit around the Sun.

Elaborate mathematical analysis of the interactions of these three effects shows that they could produce significant cycles of high-latitude cooling. The patterns created by this theoretical model closely match the estimations of changing temperature that have been gained by study of features such as the

oxygen isotope ratios in seafloor sediments whose age is known. Using these estimations, it has become clear recently that the cycles of advance and retreat of ice over the past several hundred thousand years show certain evidence of Milankovitch's 20,000-, 41,000- and 100,000-year cycles.

Cycles such as these have been found in the isotopic spoor of past climatic changes in sediments 8 million years old. But because there are puzzling variations in the patterns, it would be foolish to assume that astronomical cycles tell the whole story of the ice ages.

Other theories put forward to explain

the march of the ice ages include the suggestion that clusters of volcanic eruptions could trigger periods of glaciation. The huge amounts of dust thrown into the atmosphere by such activity could significantly reduce the amounts of solar energy reaching the Earth's surface and so cause marked cooling.

Studies of the gases trapped in the steadily accumulated ice of existing ice caps can tell us about the natural carbon dioxide levels in the atmosphere hundreds of millennia into the past. This record of potential ancient levels of the Greenhouse Effect extends through many ice ages and interglacials. The resulting correlations are intriguing, if not terrifying.

It appears that at the coldest period of the last ice age, 20,000 years ago, the carbon dioxide level in the atmosphere was much lower than it was even 200 years ago, before the industrial use of fossil fuels pushed up the level artificially. This low level implies a low degree of greenhouse warming, and suggests that the Milankovitch factors are actually linked with earthly weather systems by means of a carbon dioxide-based Greenhouse Effect.

We do not really understand how this linkage comes about, but some scientists think it results from variations in carbon dioxide levels induced by photosynthesis of plant plankton in the oceans, a process that depends on the amounts of sunlight available. If there is, in fact, a carbon dioxide coupling agent and thermostat built into the planet's weather machinery, we are producing frightful problems for the world if we override those controls by burning fossil fuels, despoiling forests and polluting seas.

The more we can understand the mysteries of natural climatic changes in previous ages, the better placed we shall be to predict changes in the uncertain, human-influenced future.

The unquiet Earth

Every aspect of the earth we live on is changing, day by day and month by month. Some changes are imperceptible—others dramatic and often devastating. Seemingly stable continents are in fact slowly moving and, perhaps, starting to fragment. More than 800 volcanoes puncture the planet's crust, many of them forming a "Ring of Fire" around the Pacific Ocean. Steam and scalding water gush fearsomely from the Earth's depths, earthquakes make the planet tremble, and giant waves may bring catastrophe to Pacific islands. The phenomena are familiar—if only from newspaper headlines. Yet many questions remain unanswered. Why, for example, is there such an uneven distribution of land between Northern and Southern Hemispheres? What actually causes volcanoes to erupt? Why should one of the coldest countries in the world—Iceland—experience so much geothermal activity?

Earth scientists seeking explanations for these mysteries are indebted to two major intellectual convulsions that radically reshaped their worldview. The first took place about 200 years ago and laid the foundations of modern geology. The second began in the 1960s, and its shock waves are still reverberating around the discipline. This is the theory of plate tectonics—popularly known as continental drift—which has provided such a rich and persuasive set of explanations for several hitherto inexplicable geological phenomena.

The first wave of revolutionary thought to hit the Earth sciences came from the Scot, James Hutton (1726–97). He was the first to propose that our planet's geological working is a continual process. In other words, the processes that we now see taking place have always happened and they are, indeed, sufficient to explain the architecture of the world we can observe.

Hutton's theories—first published in 1785—are

Folded rocks give undeniable proof of the Earth's movement.

judged by most historians to be the bedrock of modern Earth sciences. He grasped the fact that continents are eroded away, washed into the oceans and layered into the raw material of new sedimentary rock. He reasoned that under the influence of the Earth's tremendous heat, old sedimentary rock could be uplifted and twisted into new mountains and continents. And he put forward the hypothesis that, at great depths below the Earth's crust, rocks themselves could melt and erupt, through volcanoes, to form yet another source of erodable rock. His insights, which remain unchanged in essence, were quite staggering. Not only did he arrive at his conclusions with very little real evidence but he was flying in the face of established beliefs.

The plate tectonic revolution brought a new, dynamic dimension to the worldview of Huttonian geology. The theory that the surface of the Earth is divided into a pattern of ''plates'', which are slowly moving in different directions, was finally accepted in the late 1950s. As early as 1620, however, Francis Bacon, the English writer and philosopher, had remarked on the extraordinary mirror-image of the east coast of South America and the west coast of Africa. But it was the German Alfred Wegener (1880–1930) who, between 1912 and 1915, put forward the first detailed scientific case for ''drifting continents''. His theories were greeted with derision and only finally vindicated when crucial modern evidence based on the irrefutable fact of paleomagnetism proved their correctness.

These theories go much, but not all, of the way to explaining some of the Earth's great mysteries.

Continents
on the move

The Earth we live on is continuously and inexorably altering. The zones of earthquakes and volcanic eruptions that surround the Pacific Ocean are destructive but minor expressions of this fundamental truth. We cannot easily detect the changes, but clear evidence for them lies in the mountains and the oceans, as well as in the depths of the Earth itself. This evidence demonstrates the mysterious and irresistible way the continents themselves have grown, moved, split apart and come together.

It was suggested as long ago as the 1600s that some of today's continents were once joined together. But although early scientists recognized that the outlines of Africa and South America made an almost perfect fit, they could not understand how the two landmasses had been forced apart. In 1801, the German geographer Alexander von Humboldt proposed that the Atlantic Ocean had resulted from a great current scouring out the land between them.

It was not until early this century that the German Alfred Wegener (1870–1930) put forward the idea of "continental drift", now known as plate tectonics. Wegener was an astronomer and meteorologist and, despite the wide-ranging evidence he produced to support his theories, they were greeted with derision by many geologists when they were widely published during the 1920s.

Little further attention was paid to the idea of plate tectonics until the 1960s, when interest in the Earth sciences burgeoned. The existence of paleomagnetism—magnetism remaining in the rocks from past times—was discovered, and, at last, a source for the colossal energy needed to raft whole continents across the face of the globe was identified.

That energy lies beneath our feet. Deep within the Earth, the breakdown,

An awesome reminder of the slow, relentless power of Earth's crustal movements persists in the almost vertical strata of rocks in the Swiss Alps.

Some 120 million years ago, the huge northward-rafting African Plate clashed against the Eurasian Plate. As the two great landmasses met, the rocks crumpled. Sediments which had previously lain flat began to fold back and forth upon themselves. Twenty-five million years ago, these forces, generated, deep within the Earth, thrust the rocks up into jagged peaks—beautiful and deceptively unchanging to look at.

The dynamic Earth

The basic theory of plate tectonics is that the crust of the Earth is made up of six major and several minor rigid plates, which float on a weaker and more pliant layer of the mantle. They are moved about by convection currents deep within the mantle. Each of the plates moves as a distinct entity in relation to all the others, and the movement of one plate necessitates movement in those on its boundaries. Where the plates meet are zones of intense seismic activity: mountain building, earthquakes and volcanoes.

Plates are constantly being created along the mid-oceanic ridges, and are as constantly consumed at the trenches, most of which lie around the edges of the oceans. In consequence, although the oceans have existed for billions of years, no sediments more than 160 million years old have yet been found under the sea, although some rocks on land are known to date back just under 4,000 million years. Cores of rock brought up by deep-sea drilling operations can be dated by the fossils they contain.

Since the Earth's crust is not static, the shape of the world's continents 50 million years from now will have changed. On current evidence, as the Americas continue to move westward, the Atlantic will widen at the expense of the Pacific, and a long sliver of the Pacific coast of America will become an island and move farther north. South America, too, will move north, compressing Central America and Panama.

Australia and New Guinea, drifting north, will eventually collide with Southeast Asia. And as a result of Africa rafting north into Europe, the Mediterranean Sea will dry up and a new mountain chain will be raised, from northwest Africa to Turkey. The Great Rift Valley may extend, and eventually a fragment of East Africa may partially split off and move toward Asia.

The Earth's crust is divided up into a number of tectonic plates, most of which consist of both sea floor and continental landmass. The plates are in constant movement, building along the ocean ridges and being consumed at the trenches. In addition, they are all slowly drifting away from their present positions.

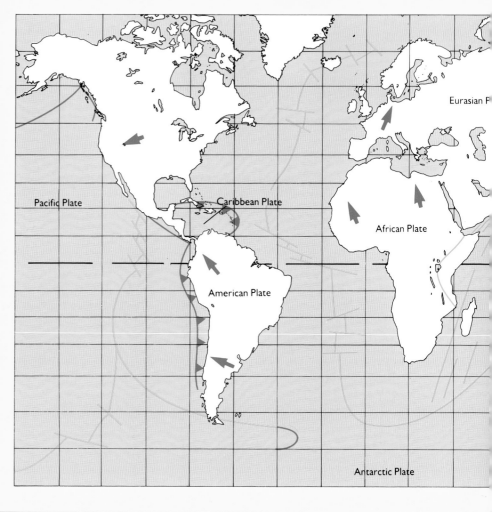

Pacific Plate

Caribbean Plate

American Plate

Eurasian P

African Plate

Antarctic Plate

or "decay", of radioactive minerals, such as uranium, thorium, and potassium, produces so much heat that the rocks become molten, as an erupting volcano dramatically shows. The heat that is constantly being produced can be lost only through the Earth's surface. Hot material rises from the deep layers to near the surface and then returns, cooled, into the depths. This continual churning of the Earth has resulted in a layered structure, and this provides

another key with which to unlock the secrets of the moving continents.

The heaviest substances now form a dense "core" 7,000km/4,350mls in diameter, which makes up practically one-third of the Earth's mass but only a sixth of its volume. Surrounding the core lies the lighter "mantle", some 2,900km/1,800mls deep. The outermost layer is made up of a thin skin or "crust"—comparable to the skin on a peach—of the lightest materials, which

float on the underlying mantle. Where it forms the ocean floors, the crust is only 5km/3mls thick, but the continents vary from 35km/22mls to more than 50km/31mls in depth where the roots of the great mountain chains project downward into the mantle.

Because the continents are thick, most of the heat produced within the Earth is lost through the thinner crust that forms the floors of the oceans. Along vast stretches of the ocean floor,

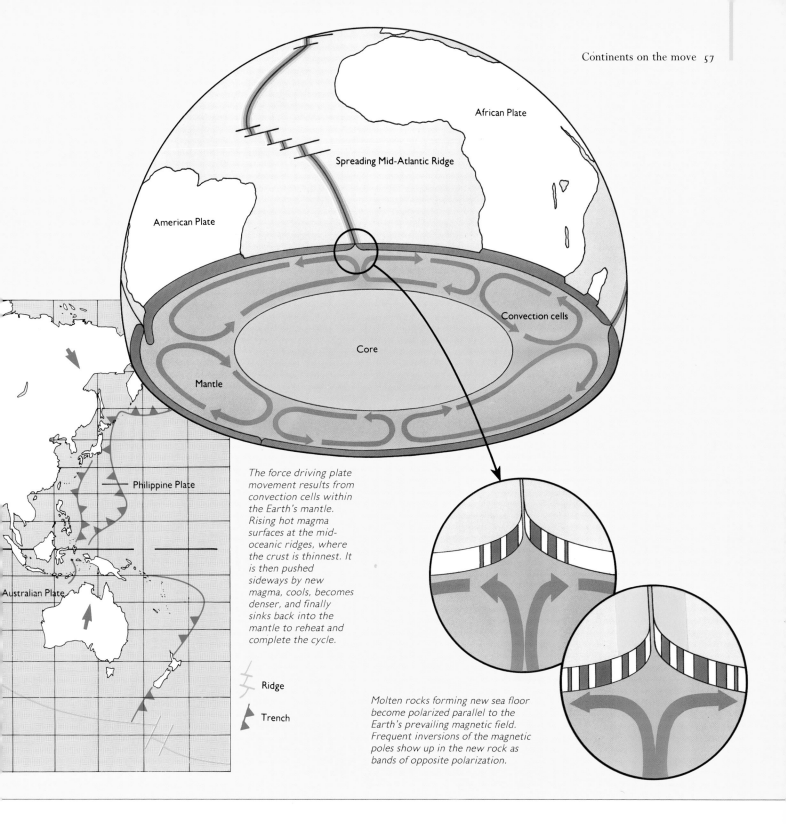

African Plate

Spreading Mid-Atlantic Ridge

American Plate

Convection cells

Core

Mantle

Philippine Plate

Australian Plate

The force driving plate movement results from convection cells within the Earth's mantle. Rising hot magma surfaces at the mid-oceanic ridges, where the crust is thinnest. It is then pushed sideways by new magma, cools, becomes denser, and finally sinks back into the mantle to reheat and complete the cycle.

⚡ Ridge

▼ Trench

Molten rocks forming new sea floor become polarized parallel to the Earth's prevailing magnetic field. Frequent inversions of the magnetic poles show up in the new rock as bands of opposite polarization.

heated material arrives beneath the crust, and volcanic chains, or ''spreading ridges'', rise within the depths of the ocean. The heated material now moves horizontally away from both sides of the ridge. As it does so, it pulls the ocean floor with it. So, along the spreading ridge lies a cleft, which would continually widen if it were not also steadily being filled with new, molten rock. Underwater photographs dramatically show these rocks, incandescent and

bubbling, as the icy waters of the ocean depths quickly cool them.

When the mantle rocks have moved away from the spreading ridge and gradually cooled, they return to the deep layers of the mantle, carrying with them the overlying oceanic crust. This subduction takes place along great trenches, such as the Marianas Trench.

If, on a map, we progressively remove the ocean floor, starting with that produced most recently, and move the

continents back to the positions they occupied before these rocks forced them apart, we find that the continents come together to form a single supercontinent. Known as Pangaea, it existed until around 200 million years ago. The northern part of the supercontinent, known as Laurasia, was made up of North America and Eurasia. The southern part, Gondwana, was composed of Australia, South America, Africa, Antarctica and India, which later broke

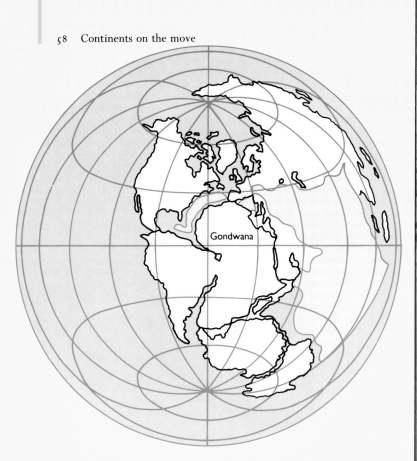

Although still joined to form Gondwana, 160 million years ago the southern continents were breaking away from those in the north, and the intervening oceans were beginning to form.

Fossil plants are important evidence of the existence of Gondwana. Indelible imprints of the tree-fern Glossopteris are widely distributed in rocks in present-day continents of the southern hemisphere. This leaf fossil, found in New South Wales, Australia, dates from the Permian period, approximately 260 million years ago.

away and moved northward to join Asia.

Why did this supercontinent break up? Probably because it was so large that parts of it lay over regions where hot material was rising from the mantle and spreading sideways. This force slowly pulled the supercontinent apart like an exploding jigsaw puzzle.

During these movements, as South America moved westward, old ocean floor to the west of the continent was consumed in a Pacific Ocean trench. But, eventually, the South American continent itself, which was too massive and light to be swallowed up, reached the edge of the trench. Further movement of the continent could now take place only by the trench consuming old sea floor of the Pacific Plate to the west. As that material slid down under South America, its movement caused instability and mountain building along the whole of the continent's western margin. And so the Andes, with the earthquakes and volcanoes indicative of their origins, were born.

Other mountain ranges have risen for different reasons as the fragments of old supercontinents collided with one another. India is still drifting northward, raising the lofty Himalayan mountain range. Africa also is moving north, crumpling up the southern parts of

The San Andreas Fault, California, is a classic example of a transform fault, where two plates slide past each other. No building or subduction takes place, but earthquakes are common.

Europe into mountains such as the Alps. The Pyrenees have risen as Spain has rotated counterclockwise into France.

All this is no new process. Asia was once a separate continent, and its collision with Europe about 250 million years ago caused a chain of high mountains, whose eroded stumps now form the Urals. These events must have changed the climates of the different regions as well as their geography, as new deserts, cut off from rain-bearing clouds, grew in the lee of the mountains.

The "frozen magnetism" in the rocks allows us to trace the history of the continents back to a time before the formation of the oldest sea floor. And it is now clear that India is not the only fragment of Gondwana that split away and drifted north to collide with Asia. Tibet, Southeast Asia and China were all once separate from Siberia and have collided with it at different times from 350 to 250 million years ago.

Even with such knowledge, we tend to assume that our world is stable and unchanging, until an earthquake or volcanic eruption along the fringe of one of the tectonic plates brings home to us the arbitrary and violent results of their relentless movement. Suddenly, we are chillingly reminded of the awesome and mysterious power of the restless Earth.

Forges of
the fire god

The processes of great power that shape our planet's surface take place so slowly that they do not normally impinge on human perceptions. So great is the time scale involved that if the Earth's geological history could somehow be captured on film, we would see no more than a single frame in a whole lifetime. But there are moments when geological change occurs with terrifying suddenness, and when unimaginable forces are unleashed with cataclysmic results.

Many miles beneath the surface of the Earth, a vast reserve of heat energy remains stored in molten rock, or magma. When this magma meets a weakness in the Earth's crust, it breaks through to the surface, and the result is a volcanic eruption.

Volcanoes are powered by giant magma chambers, held in place by the tremendous pressure of the rock above and around them. The magma is lighter than the surrounding rock, and finds its way upward through the smallest of fissures. As it rises, the gases dissolved in it expand with such great force that eventually the gas and magma together blast a hole through the Earth's crust. The gas pours into the atmosphere, carrying dust and ash skyward, while the magma flows over the ground as lava.

This disgorging of the Earth's primeval fire, a phenomenon that has long mystified and intrigued the planet's human inhabitants, is not always an isolated, catastrophic event. Some types of magma form freely flowing lava when they reach the surface, and this rapidly runs away down any incline, cooling and solidifying. Because the energy in the magma is slowly released, volcanoes with this type of lava are often active for centuries without cataclysmic eruptions.

Other kinds of magma have a viscous, almost toffeelike consistency. The cooling lava piles up in collapsing columns

Magnificent, intriguing yet terrifying, a relentless flow of lava pours constantly from Hawaii's great volcanoes. Islanders have learned to live alongside these burning streams, but familiarity has not bred contempt, for no one knows how magma—the stuff of lava—is formed inside the Earth, where it comes from, nor when it might burst forth, to overwhelm their world in seconds.

and jagged blocks, forming volcanoes with high, steep-sided cones. As long as the flow of magma is maintained, the forces beneath the crust are safely released. But if the flow becomes blocked by solidified lava above the magma chamber, the situation is very different.

When such a blockage occurs, the forces urging the magma upward continue to act and enormous pressures build up, sometimes over decades or even centuries. On the surface, changing ground levels or escaping gas may hint at the build-up of pressure beneath, but often there is no indication that anything is about to happen. This lack of activity, in itself, is ominous.

A period of threatening silence was exactly what heralded one of the great eruptions in modern history. In 1883, the Indonesian island of Krakatoa had shown no signs of volcanic activity for two centuries, and its slopes were clad in thick forest reaching down to the warm waters of the strait that divides Sumatra from Java. But beneath this busy shipping lane a gigantic volcanic bomb had been primed—one whose effects were to be felt worldwide.

In May 1883, Krakatoa burst into life. A colossal explosion was heard more than 160km/100mls away, and a huge column of ash and dust was borne aloft by a blast of air heated to furnacelike temperatures. This initial eruption was followed by bursts of activity over the next two months but these were a mere prelude to the final spectacle.

The word "paroxysm" is used by vulcanologists to describe the cataclysmic release of volcanic energy. On 26 August, Krakatoa was gripped by one of the greatest paroxysms ever recorded. A column of black ash raced toward the upper atmosphere, plunging the strait into darkness. The volcano partially collapsed, triggering off a tsunami—a giant wave—that claimed the volcano's first victims on mainland Java. Throughout the night, there were further eruptions, which could be heard more than 1,000km/620mls away.

The final act in the island's self-destruction took place during the following day in four explosions, the third of which was probably the greatest in the whole of recorded history: nearly 20cu km/4.75cu mls of ash was thrown into the air. Of the 36,000 people who died, those who were not asphyxiated by dust or poisoned by toxic gases were drowned in tidal waves.

The after-effects of these colossal explosions were felt all around the world. Atmospheric shock waves emanating from Krakatoa crossed and re-crossed the globe, while as far away as North America volcanic dust, swept up into the stratosphere, created abnormally brilliant sunsets. Strangest of all, floating pumice, a lightweight, glassy rock, clogged the seaways and beaches all around the Indian Ocean. Krakatoa itself had all but vanished beneath the sea, leaving only a gigantic flooded crater surrounded by fragments.

There are probably about 850 active volcanoes on Earth, of which nearly a quarter are under the seas. They are not randomly scattered, but are clustered in

Cone built up by ash and lava from successive eruptions

Volcanic coastal mountains, such as the Cascades and the Andes, are formed above a rich source of magma by an oceanic plate sliding beneath a lighter, thicker continental one.

Molten intrusion

Upper mantle

Partial melting

Continental crust

Trench

Oceanic crust

Types of volcanic eruption

Peléean eruption
Named for Mt Pelée, Martinique. Blocks of viscous lava are ejected, along with a nuée ardente, a "burning cloud" of hot volcanic gases.

The church tower stands as a monument to the buried San Juan

Courting disaster

Early farmers discovered that volcanic soil was extremely fertile and settlements soon grew up on volcanoes' slopes. A high price has often been paid for this temerity. In AD79, Pompeii and Herculaneum were swallowed in ash and pumice from Vesuvius, and a deluge of ash and toxic gas destroyed St Pierre, on Martinique in 1902.

The village of San Juan Parangaricutiro in Mexico was, atypically, buried slowly. In February 1943, a huge rift opened up, out of which boiled red-hot lava and rock. The volcano, known as Paricutín, took a week to reach 168m/550ft; a year later it was 336m/1,100ft high, and ash and lava covered the land for 32km/20mls around.

Hawaiian eruption
Moderately violent but long-lived, with a continuous discharge of freely flowing lava, often through secondary fissures in the volcano's side.

Strombolian eruption
Typical of the Italian volcano. Small bombs of sticky lava, gas and cinders are spasmodically thrown a short distance into the air.

Vulcanian eruption
Named for Vulcano, near Stromboli. Rare explosions of very viscous magma hurl large glowing blocks a considerable distance.

Plinian eruption
Cinders and ash under immense pressure are ejected high into the air. Pliny saw and described the eruption of Vesuvius in AD79.

The cataclysmic eruption in 1980 of Mount St Helens in the Cascade Range in Washington State, USA, was the best documented volcanic explosion ever to have taken place.

In March, a succession of earthquakes in the mountain's vicinity had convinced scientists that an explosion threatened, and by 10 May an already ominous bulge of some 91m/300ft on the north face was growing by 1.5m/5ft a day. In the opinion of David Johnston, an American geologist who lost his life in the eruption, the mountain was a powder keg, with the fuse lit, but no one knew how long the fuse was.

When the explosion finally came at 8.32am on 18 May, about 4cu km/1cu ml of the mountain's summit was blown into the sky, leaving it 400m/1,300ft short of its former height of 2,549m/8,364ft. Shattered trees lay on the ground in their millions, all

After the initial horizontal eruption of Mount St Helens, a vertical explosion forced a column of hot gas, ash and rock 19km/12mls into the atmosphere.

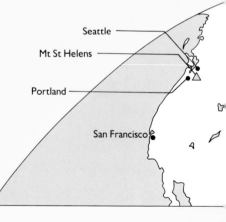

groups and chains that must have puzzled early vulcanologists attempting to rationalize these outpourings of the Earth's interior. The largest of these chains, the "Ring of Fire" that girdles the Pacific, forms a horseshoe-shaped arc of volcanic regions, which stretches from New Zealand to the Bering Sea, and from there almost to the tip of South America.

It was only earlier this century that the mystery behind the siting of these volcanic chains was finally explained by

the theory of plate tectonics. The Ring of Fire, for example, is formed where the Pacific Plate meets five other plates bearing continental landmasses. Krakatoa lies in the most intensely volcanic part of the world, where the Indo–Australian Plate is being driven beneath the Eurasian Plate, just off Indonesia. Mount St Helens, which exploded in 1980 after more than a century of silence, lies near the junction of the Juan de Fuca Plate below the North Pacific and the plate that forms North America.

And the volcanoes of the Mediterranean, which include Etna, Stromboli and Vesuvius, lie near the point where the African and Eurasian plates meet.

There are some volcanoes which seem to defy this rule, and which erupt not at plate boundaries, but far from them. The volcanoes of Hawaii, for example, lie almost in the middle of the Pacific Plate. Volcanoes such as these are created by "hot spots", or plumes of magma, that rise from a fixed place in the Earth's mantle. Why these hot spots pointing away from the mountain, and making more than 575sq km/225sq mls impassable. No vegetation remained for animals to feed on, and rivers were clogged with ash and debris, killing all life in them.

For weeks after the explosion a huge area seemed as lifeless as the Moon. Yet, despite the tremendous devastation, nature began to reassert her hold on the ash-choked slopes. Volcanic ash, unlikely though it may appear, is a highly fertile substance, charged with the minerals plants need for growth. Within months of the Mount St Helens explosion, seeds borne on the wind had germinated and bloomed. Saplings have taken root, and eventually the forest will re-establish itself. In the late 1980s the first small parties of climbers were allowed back into the area to explore and to marvel at nature's power, both to destroy and to generate life.

Another huge Plinian eruption occurred in July. The mushroom cloud formed as the original column lost energy and began to roll outward.

Thousands of acres of forest were laid waste by ash and sludge when lakes and rivers burst their banks. The tiny figures give some sense of the scale of the disaster.

should exist has not yet been entirely explained, but it is known that as the crust plate drifts overhead, the rising magma forms chains of volcanic islands.

In the Hawaiian Islands this process has created a string of volcanoes more than 600km/370mls long. The volcanoes that formed the most westerly islands have long since fallen silent, while those to the east, on Hawaii itself, directly above the hot spot, are constantly erupting. Hawaii's Mauna Kea is the largest active volcano in the world.

Volcanoes are often classed as active, dormant and extinct, but these labels can be misleading. Recorded history is too short a yardstick by which to measure the activity of a volcano. It may be active for hundreds of thousands of years, with each successive eruption being followed by centuries of apparent inactivity. Its last eruption may be before recorded history, but its apparent dormancy or extinction may well disguise preparations for an explosion.

Today the immediate likelihood of an eruption can be assessed with a battery of precise measuring instruments. While they cannot actually predict the moment of an eruption, they do allow its probability to be estimated.

In the days before vulcanology became a scientific discipline, predicting eruptions was a more haphazard business, and was based on the correct interpretation of the signs that accompany the gradual build-up of pressure beneath the Earth's surface. Sudden releases of boiling water or sulphurous gas would have acted as warning signs, as would earthquakes or changes in ground level. Yet these signs were often of little help because they came too late for the inhabitants to flee to safety.

In some eruptions, an avalanche of ash and toxic gases, known as a *nuée ardente*, can advance faster than 250kmh/150mph, engulfing all that lies in its path. It was such a nightmarish avalanche that destroyed the town of St Pierre in Martinique in 1902, killing all but two of its 30,000 inhabitants. Almost as rapid is a volcanic mudslide, known by its Indonesian name, *lahar*. Triggered when molten rock meets water, snow or ice, and mud, a lahar smashes trees and houses with its weight; and once it has come to a standstill, the mud sets, trapping its victims as if in concrete.

Our knowledge of the effects of eruptions necessarily diminishes as we go farther back in time, because without written records, the size of eruptions in the distant past is difficult to gauge. The evidence for ancient eruptions is based partly on geology, and partly on myths and legends first handed down orally.

Geological investigations certainly indicate that in 4,000BC, the explosive collapse of Mount Mazama in North America must have been one of the largest in recent prehistory. Crater Lake, which measures 10km/6mls across, is a testament to the amount of ash and rock that must have been ejected into the sky and then swept across the continent. In Nebraska, the remains of rhinoceroses buried in ash testify to an even greater explosion, probably around 10,000 years ago, but the site remains a mystery.

Mysterious, too, is the more recent explosion which is known to have shattered the Greek island of Thera, or Santorini, in about 1450BC. All that remains of Santorini today is a giant submarine crater in the Sea of Crete, 80sq km/31sq mls in area. It has been calculated that the explosive force that created this huge crater must have thrown 30cu km/7cu mls of ash skyward, blocking out the sunlight over the whole of the eastern Mediterranean for several days.

Tantalizingly, no direct records have been found of this extraordinary event. However, circumstantial evidence from the nearby island of Crete suggests that the eruption could have destroyed not only towns and cities, but also the entire, highly ordered Minoan civilization, which held sway over much of the Aegean at that time.

Archaeologists and historians have debated the reasons for the sudden collapse of the Minoan culture. Until recently, it was thought that the explosion on Santorini came too late to be a factor in this epic disaster. However, new dating evidence, obtained from sources as distant as glacial ice in Greenland and bristlecone pines found in California, suggests that the destruction of Santorini may well have brought with it the destruction of the Minoans.

It may also be that the legendary disappearance of the city of Atlantis is an account of the destruction of Santorini. And the "pillar of cloud" mentioned in the Bible, in Exodus, bears an uncanny resemblance to a distant trail of a volcanic eruption.

Threat over the Mediterranean

The great classical civilizations of the Mediterranean developed in the shadow of volcanoes. Indeed, the very word volcano names the fiery island of Vulcano, lying just north of Sicily in the Tyrrhenian Sea.

The sailors of ancient times were well used to the unearthly light from these mountains of fire, for a few miles from Vulcano is another island volcano, Stromboli; while to the south, on Sicily itself, is towering Etna. Vesuvius stands threateningly over the Bay of Naples, and eastward, in the Sea of Crete, the shattered remnants of Santorini, or Thera, mark the site of one of the greatest volcanic eruptions of the Mediterranean.

The eruptions of these smoking mountains inspired not only fear of the physical world in the people living on their slopes, but also fear of the supernatural one, and the gods were held responsible for the volcanoes' violent outpourings. Even today, in many parts of the world where volcanoes spew fire and destruction, the gods of the underworld are still feared.

Could Thera have been the lost city of Atlantis? Wall paintings, which have survived burial under some 30m/100ft of volcanic ash for 3,500 years, reveal a sophisticated society with a well-organized army, a prosperous commercial life, houses with many floors, and graceful ships—all signs that the story may be true.

A fierce and vengeful Hawaiian fire goddess, Pele is believed to live in Kilauea crater. The carved wood figure has real hair and eyes made from shells.

The ancient Greeks thought that the god Hephaistos was responsible for the outpourings from volcanoes, while the Romans ascribed eruptions to Vulcan, blacksmith to the gods, who forged thunderbolts for Jupiter in his smithy under Mount Etna. The lava thrown out by the volcano was seen as sparks flying from Vulcan's forge.

Later, Giorgio Vasari (1511–74), found in the old story a vigorous theme for his painting, Vulcan's Forge.

Energy from the underworld

As the Vikings sailed in their longships toward the coast of Iceland, a strange sight met their eyes. Plumes of dense white smoke ascended into the sky, but with no sign of flames, and no people to have built fires. They named the place Reykjavik—"Smoke Bay"—only to discover, when they later landed on the shore, that the white columns were not smoke but steam.

These spectacular displays were the outpourings of geysers, which dot the Icelandic landscape more densely than anywhere else on Earth. The original namesake for this worldwide phenomenon is a powerful jet of water and steam known as Geysir, in Haukadalur, Iceland, which once reached heights of 50m/165ft with regularity, though it now erupts only rarely. The name may originally have come from *geysa*, an Old Norse word meaning "to gush".

Iceland owes its many geysers to the fact that it sits astride the Mid-Atlantic Ridge, where two great segments of the Earth's crust are moving away from each other, wrenching the Old World from the Americas, a process that has continued for many millions of years. Where the two plates of the Earth's crust separate, they leave a thinner area below the surface, a line of weakness into which molten magma wells up, forming pockets in the spongy rock above.

From these magma chambers, the fiery liquid can flow laterally through the crust. As it does so it warms the rocks around it, and water which has soaked down through the porous rock encounters these heated rocks. What happens next depends on the exact circumstances.

If the water can escape freely to the surface, it will bubble up to form a hot spring, or a boiling mud pool. If, on the other hand, the water is enclosed by rock, it may become superheated but unable to boil because the massive pressure of the rocks above prevents steam from escaping. The water remains in its subterranean cauldron, invisible on the surface unless it happens to be beneath a glacier, in which case it may melt the ice above to form a lake.

When circumstances are somewhere between these two extremes, the outcome is far more dramatic. If the water is partially enclosed—enough to keep it from boiling though its temperature rises well above boiling point—it can build up a "head of steam", which then blasts the superheated water out of its underground confines. The result is a geyser, or, if steam alone is ejected, a fumarole.

The exact workings of most geysers remains a mystery, but in the case of Old Faithful, in Yellowstone Park, USA, geologists believe they know what is happening in the depths of the Earth below. They suspect that Old Faithful originates in a long, roughly U-shaped

"Roaring Geyser", Grothrrmal, below, at Hvrravellir in Iceland, takes its name from the deep, rumbling sound it produces before spouting scalding water and steam into the air. Minerals in the water here have built up layers of sinter to form rocky cones around some of the most spectacular hot springs in the world.

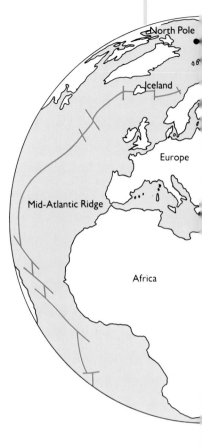

Iceland is cut in two by the Mid-Atlantic Ridge, where two great plates of the Earth's crust are pulling apart. Here, molten magma, welling up from deep in the Earth, heats underground water which escapes through geysers.

Iceland's most famous geyser is Strokkur, the "Churn". At first sight it seems nothing more than a tranquil, sparkling pond; only a few wisps of steam give a clue to its true nature. Then the water level begins to rise and fall and for an instant a dome of clear water wells up. This erupts into a pillar of steam and boiling water 30m/ 100ft high, usually at intervals of nine minutes around the clock.

tube that fills with groundwater. At the far end of the tube, at a higher level than the geyser outlet, is an empty chamber surrounded by hot rocks. Water dripping into this chamber turns to steam which gradually builds up the pressure in the chamber.

When eventually the critical pressure is reached, steam forces its way out by ejecting the water from the U-shaped tube. According to popular myth, Old Faithful erupts like clockwork once every 65 minutes, but the truth is that all geysers are somewhat fickle. Even this model of loyalty can go quiet for extended periods, while on occasions it may erupt every half hour.

The geysers and hot springs of Yellowstone mark a site of ancient volcanic activity, where the last eruption probably occurred as little as 10,000 years ago. This is a characteristic stage in the slow, lingering death of a volcano, known as the "solfatara" stage. The name comes from the village of Solfatara, near Pozzuoli in Italy, which was famous in antiquity for its hot mud pools and springs. But the fate of this little village shows that the solfatara stage offers no certainty that a volcano is in terminal decline. In 1538, Solfatara itself was engulfed by a colossal cindercone eruption

Hot, pressurized water is a powerful solvent, and as it forces its way through the Earth's crust it is constantly eating away at the surrounding rock. It emerges at the surface laden with dissolved minerals, which are deposited the moment the water cools. Exactly what minerals the water contains will depend on the nature of the rock.

Sometimes the minerals brought forth from the Earth by a hot spring are radioactive, and they impart this quality to the water or mud. The supposed therapeutic potential of such springs and mud pools may be due to this, or to the minerals in the water, or simply to the heat itself, which can be soothing and beneficial in diseases such as arthritis.

Pammukale, in Turkey, became a health resort as early as 190BC, when Eumenes II, King of Pergamum, founded the holy city of Hierapolis. It thrived under the Greeks and later the Romans, and was renowned for its temples and sacred hot pools. One part of Hierapolis, however, was anything but healthy; this was the Plutonium. It is still there, a small cave where a noxious mist hangs over the steaming rivulet within.

The gases of the Plutonium are so powerful that they are reputed to kill any animal within seconds. Such gases are an occasional by-product of the underground alchemy of hot springs, which may dissolve different types of rock and mix the products together in a potent, reactive brew.

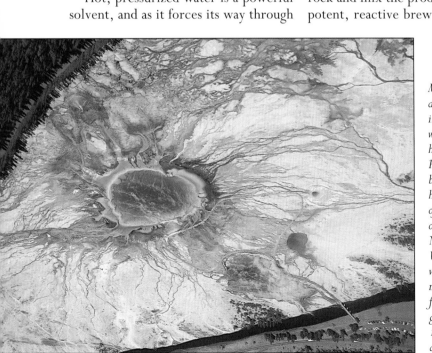

Microscopic bacteria and algae that thrive in the near boiling water encircle the heart-shaped Grand Prismatic Spring with bands of brilliant hue. Under the clouds of steam that swirl over Yellowstone National Park, Wyoming, lies the world's largest and most varied geyser field, with some 300 geysers and almost 10,000 hot springs and fumaroles.

Harnessing Earth's energy

The heat of the Earth's interior is a vast energy reserve, one that dwarfs the potential of all the world's fossil fuels. Until the beginning of the twentieth century, this resource was largely unexploited. Then, in 1904, the world's first geothermal power plant opened at Larderello in Italy, and since then increasing use has been made of this virtually pollution-free source of power and heat.

The most energy-rich geothermal sites are those where the furnacelike heat below ground produces high temperature steam, which forces its way to the surface. At the Geysers geothermal site in California, natural steam generates enough electricity to provide power for half of San Francisco.

In the mountains of northern Japan, the macaque monkeys have discovered the delights of a dip in a thermal spring when the temperature drops below freezing. While snow storms rage, they keep warm by sitting in the hot water.

Without the natural bounty of geothermal energy, life in Iceland would be hard and bleak. But by tapping this resource and creating geothermal plants, such as this one at Svartseng, to harness the heat and steam, people can enjoy all the comforts of modern life. They even have the blessing of year-round outdoor swimming in hot, clear blue water.

Engineers plan to develop power plants in other parts of the world where hot rock lies near the surface, but where steam does not normally form. This can be done by injecting water into the rock through a borehole. The water boils on contact with the rock, and the steam produced emerges through a second borehole and is used to drive turbines.

In some places, such as Iceland, geothermal energy produces large quantities of hot water. This cannot easily be used for generating electricity, but it is valuable for heating. With the benefit of almost unlimited heat at the turn of a tap, Icelanders are able to enjoy not only warm homes but also fruit and vegetables grown in geothermally-heated greenhouses.

Shockwaves and fatal flaws

In 1835, a remarkable event took place along the coast of southern Chile. In a single convulsive movement, the entire shoreline rose and came to rest a yard or more above its previous position. Ships seemed to sink suddenly as quaysides lurched upward, and sandbanks that had been submerged became low islands. The uplift reached 3m/10ft in places, and beds of barnacles and mussels were carried high out of the water.

The shock waves that accompanied this upheaval caused tremendous devastation. In the port of Valdivia, the boards of wooden houses burst open, and the water in the harbour rose and fell abruptly, creating powerful eddies that pulled boats from their moorings.

We know much about these events because they were witnessed by Charles Darwin (1809–82), who, on 20 February, had disembarked from the survey ship *Beagle*. The earthquake struck as he was exploring a wooded hillside nearby. He felt not only immediate fear but also a profound sense of shock that lingered for many days. It is the realization that the Earth—the "emblem of solidity",

as Darwin pithily described it—can suddenly writhe and shudder that makes the experience so terrifying. It shatters one of our most fundamental preconceptions, giving earthquakes their peculiar and special menace.

Earthquakes, like volcanic eruptions, do not occur at random across the Earth's surface. In some parts of the world, minor tremors are a familiar and often unremarked part of daily life. In these same areas, major earthquakes are, however, an ever-present threat. Over the centuries, they have been woven into lore and legend, and today still strike fear into the human psyche.

Our search for an explanation for earthquakes extends far back beyond recorded history. Innumerable myths have evolved, many of which share a common theme. The central belief of such legends is that the Earth is supported on the back of some large animal, and that the restlessness of this giant creature sets off earthquakes.

In South America, Darwin heard a different explanation for the 1835 earthquake, one that showed an inkling of

Most earthquakes occur on margins of tectonic plates, either where they slide past each other or meet head-on, as in the Himalaya Mountains.

A shattered home in San Francisco, built astride the San Andreas Fault, bears witness to the unpredictable power of earthquakes.

understanding of the real forces at work. The earthquake had been caused, so the story ran, by an old woman, who had mysteriously sealed the crater of a nearby volcano. The silenced volcano had only one way to vent its fury—by shaking the ground instead of pouring forth fire. The story contained a germ of truth, and it may have arisen from the observation that volcanic activity often undergoes sudden changes just before an earthquake occurs.

Since the dawn of recorded time, humans have attempted to fit earthquakes into some sort of pattern, so that warnings could be given of impending disaster. The portents that were supposed to herald earthquakes were many and varied. They included not only ominous astrological signs and inauspicious phases of the Moon, but also sudden fluctuations in the water level in wells, unexplained lights playing over the ground, evil-smelling air, strange cloud formations and unusual reactions on the part of animals.

While modern seismologists would dismiss the idea that stars millions of light-years distant could have any bearing on movements of the Earth's crust, other time-honoured portents are now the subject of detailed investigation. Some seismologists have suggested that the gravitational pull of the Moon plays a subtle part in triggering earthquakes. Changes in the water table may echo movements deep in the crust, and escaping sulphurous gases and electrical disturbances may accompany them. Many well-documented instances exist of animals fleeing before earthquakes, and it may be that a keen sensitivity to vibration alerts them to danger.

Despite centuries of theorizing and

Energy from earthquakes travels in the form of waves. Primary (P) waves compress and expand the rock and can pass through fluids, including the molten core of the Earth.

Secondary (S) waves can only pass through rock, and shake it with an up-and-down or side-to-side motion.

observation, earthquake prediction remains an art in its infancy. Some lives have been saved by short-term prediction, but earthquake forecasts that extend over periods of months or years remain too generalized to be of much immediate value.

Earthquake measurement, on the other hand, has now been perfected to a high degree. Since the Chinese invented the first seismographs, instruments have been developed that enable the intensity of an earthquake to be measured with great accuracy, and its origin to be precisely pinpointed. A number of scales of earthquake intensity are now used, the one devised by Charles F. Richter (1900–85) being the best known.

The question of why some parts of the world are singled out for earthquakes while others are spared prompted the nineteenth-century Irish engineer, Robert Mallet (1810–81), to draw up a map of the world's earthquake areas. Although he was not in a position to know it, his map shows with great clarity that these areas, like those of

volcanoes, are concentrated along the boundaries between Earth's crustal plates. We now know that it is the collision or separation of these plates that creates shock waves, which in turn make an earthquake.

Most earthquakes go completely unnoticed. Apart from the multitude of minor tremors of the Earth's crust, many occur in uninhabited areas or under the sea, where new sections of crust are constantly being formed as the plates pull away from each other. Here quakes are common; but because they emanate from near the surface, in relatively flexible rock, they are usually weak. Even so, submarine earthquakes are sometimes felt aboard ships, where

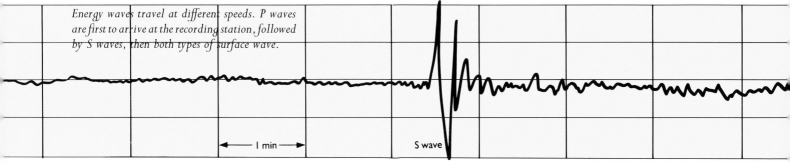

Energy waves travel at different speeds. P waves are first to arrive at the recording station, followed by S waves, then both types of surface wave.

← 1 min →

S wave

A switchback of fractured steel girders and buckled rails—all that remains of a bridge on the Nagara-Gawa rail line after a disastrous earthquake. Japan, one of the most earthquake-prone countries in the world, lies in a subduction zone, where the Philippines Plate and the Pacific Plate disappear beneath the huge Eurasian Plate.

Love waves, left, and Rayleigh waves, below, are both surface waves. They create a side-to-side and a rolling motion respectively.

the shock waves may produce a sharp jolting movement.

Far more powerful are earthquakes caused when crustal plates collide. Although the relative movement of adjacent plates may be no more than an inch or so a year, the energy of this movement is colossal. If gradual movement is prevented by frictional forces locking the plates together, then stress builds up until it is strong enough to

P wave   Surface   wave

overcome the force of friction. In a sudden, jarring movement, a massive release of energy—many times more powerful than the largest atomic bomb—sends shock waves racing through the surrounding rock at speeds up to 800kmh/500mph.

The world's strongest earthquakes occur where one crustal plate is being "subducted", or forced beneath another. At one of these plate boundaries, earthquakes may be triggered as much as 700km/435mls beneath the Earth's surface, where the two sheets of rock making up each plate collide with an imperceptible but unstoppable movement.

Deep earthquakes like this occur mainly around the margins of the Pacific, where the plates making up the floor of the ocean are being driven beneath those that bear continents and chains of islands. Sometimes, movements deep in the crust are echoed at the surface in a dramatic way. In 1923, for example, the sea floor off part of the Japanese coast fell by more than 100m/330ft.

So great are the forces unleashed by these deep quakes that the shock waves can cause havoc at the surface. In human terms, the greatest damage sometimes occurs not at the earthquake's epicenter—at the point on the surface directly above its origin—but some distance from it.

In 1985, the Cocos Plate, under the floor of the Pacific Ocean, lurched forward and under the adjacent North American Plate, just off the coast of Mexico. The earthquake occurred many miles below the coast. As with all earthquakes, its shock waves moved outward in all directions—not only straight up toward the epicenter, but also down and sideways.

It was the sideways-moving shock waves that had such catastrophic results for the inhabitants of central Mexico City, 400km/250mls northeast of the earthquake's point of origin. Mexico City, founded in Aztec times, was built on raised ground in the middle of a lake, but over centuries the waters of the lake dwindled until only silt remained. When the shock waves arrived, the silt

resonated like a drumskin. So violent was the shaking that hotels, office blocks and hospitals were brought to the ground, killing more than 10,000 people.

In some parts of the world, the plates of the Earth's crust do not meet head-on, but instead scrape past each other, grinding rock into powder and leaving linear faults or gashes on the Earth's surface. Hundreds of these faults scar the Earth, but undoubtedly the most notorious is the San Andreas Fault, which slices through California.

The San Andreas fault marks the point where the Pacific Plate meets the North American Plate. The Pacific Plate not only lies under the sea floor, it also carries with it part of California's coastline. Bound together as one unit, sea floor and coast are moving northward at the rate of about 5cm/2in a year.

In some places, the movement is a creeping one. The coast and the hinterland slide uneasily past each other, generating many small shocks, known as "strike slips". At the boundary between the two moving landmasses, strange effects show where the ground is being torn apart. Streams and rivers turn sudden right angles where their beds are diverted by the fault and old fences become twisted and buckled.

But in other places, the plates lock together. For years, no movement occurs and then the Pacific Plate catches

Japan legend has that earthquakes e caused by a giant tfish that lives half-uried in mud at the ottom of the sea. his namazu is nned down by a eat stone with agic properties and watched over by the ashima god. But as on as his back is rned, the namazu dulges in ischevious activity. aving recaptured e wayward creature, e god is shown here dmonishing morseful small amazu, who present past quakes.

Measuring earthquakes

Predicting earthquakes is still largely haphazard, although modern technology has helped seismologists—scientists who study earthquakes—gradually to build up their knowledge and understanding of these often catastrophic convulsions of our planet. It has been simpler to establish methods of measuring the intensity of earthquakes—how much energy they release.

Most commonly used is the Richter Scale, devised by American Charles F. Richter in the 1930s. This scale is logarithmic, that is, a step of one magnitude is an increase of ten times. The strongest earthquake ever measured by it, in Japan in 1933, registered just under 9; the San Francisco earthquake of 1989 measured approximately 7, only one-hundredth of this amount.

A second system of measurement, the Mercalli Scale, with steps from I to XII, measures earthquakes in terms of their effects. Tremors of intensity IV are felt by almost everyone indoors; those of intensity VIII cause the collapse of some buildings, while an earthquake of intensity XII wreaks complete devastation.

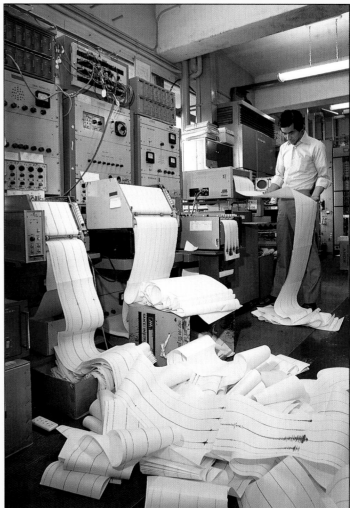

In high-risk areas, such as Japan, modern seismologists monitor the Earth's movements constantly. Batteries of sensitive instruments produce printouts of seismic waves which reflect even the slightest tremor.

The first attempt to detect distant earthquakes was made in the second century AD by the Chinese astronomer Chang Heng. This copy of his ingenious device shows dragons with balls delicately poised in their mouths. Vibration of the Earth would cause the appropriate ball to fall into the gaping mouth of the waiting frog, so indicating the direction of the quake.

In 1906, mysterious subterranean grinding of Earth's crustal plates along the San Andreas Fault brought hundreds of buildings, including the Town Hall, left, crashing down in San Francisco. Later, fire swept through the city, causing total devastation.

During the 1989 earthquake, it was the old-style houses that suffered. Modern skyscrapers designed to resist earthquake movement, such as the 260-m/853-ft pyramidal Transamerica Building, with its triangular supports and steel columns at the base, came through the earthquake unscathed.

up on lost time, effortlessly tearing apart buildings and bridges as it surges forward by up to 10m/33ft in a single devastating jump. One of the largest of these occurred with catastrophic effect in 1906, when San Francisco was brought to the ground by an earthquake and then swept by fire. For most of this century, Californians have awaited the next big earthquake.

In 1989, the earthquake that affected San Francisco caused billions of dollars worth of damage, as houses and elevated highways collapsed. Despite its force—7.1 on the Richter Scale—fewer than 300 people died; but this earthquake was not the long-expected "Big One". If this major movement of the fault occurs near one of California's cities, the toll could well be far higher.

When an earthquake strikes, those in its grip feel an intense urge to escape into the open. And well they might, because even a violent earthquake is relatively harmless out of doors. It is only when an earthquake strikes a built-up area that it becomes truly disastrous. Buildings may offer sanctuary against the elements, but unless specially designed, they can become death-traps.

As the world's human population has grown, more and more cities and towns have been built in earthquake areas, sometimes with disastrous results. One

of the most catastrophic earthquakes in Europe, for example, took place near Lisbon, which happens to lie close to the point where the Eurasian Plate meets the African Plate to the south.

In 1755, Lisbon was the crowded and prosperous capital of an empire that stretched across the Atlantic to Brazil. On 1 November, All Saints' Day, when most of the city's inhabitants were in church, the earthquake struck. The first shocks sent the terrified congregations fleeing for safety outside, but so narrow were the streets that they offered no escape. Within minutes, subsequent shocks brought whole buildings crashing to the ground.

Giant waves reverberated over the estuary of the Tagus and swamped the quaysides, carrying away people who had sought refuge at the water's edge. To complete the catastrophe, fire broke out and raged over the ruins, adding to the death toll which reached 50,000.

In the three centuries since the destruction of Lisbon, a number of the world's major cities have been singled out for devastation by earthquake. Miraculously, only about 700 people died in the San Francisco earthquake of 1906, but more than 100,000 lives were lost in Tokyo in 1923, when again a major city was first levelled by an earthquake and then swept by fire. The houses were

built of lightweight materials to make them safer in the event of an earthquake, but this had the side-effect of making them highly inflammable. Throughout the city, cooking stoves toppled over as the earthquake struck, and whole areas disappeared in flames.

But even this loss of life was dwarfed by a "rogue" earthquake which hit the Chinese city of Tangshan in 1976. Lying in eastern China between Beijing (Peking) and the Yellow Sea, Tangshan, with a population of nearly a million people, had no previous record of earthquake activity. But early in the morning of 28 July, at a time when most people were asleep, an earthquake levelled the city's flimsy houses, apartment blocks and factories. The death toll was reported to be nearly three-quarters of a million, but three years later this was revised to

250,000 by the Chinese authorities. The exact toll will probably never be known.

It has been calculated that by the year 2000 more than 100 of the world's cities will have populations of two million people or more, and of these at least a dozen—the "supercities"—will have populations exceeding 10 million. More than two-fifths of these cities, including Mexico City, Tokyo, Beijing and Los Angeles, are near plate boundaries, where earthquake threat is greatest.

There is no way in which earthquakes can be prevented, or cities moved. The only way in which disaster can be averted is by making buildings safer and improving our ability to predict these cataclysmic convulsions of the Earth's crust. The years of the next millennium will show how well modern man can adapt to this age-old danger.

The central area of Mexico City is built on an old lake bed, which has silted up and settled over the centuries. When an earthquake struck in September 1985, the vibrations in the silt layers were at the same frequency as those of the conventionally built high-rise buildings, and more than 1,000 collapsed. The city's future safety depends on rebuilding to designs that will withstand earthquake shock and movement.

Tsunami: the great wave

It was a calm evening in November 1837, and there was little about Kahului beach, on the Hawaiian island of Maui, to suggest any impending disaster. Then the sea began to retreat from the bay, silently but with great speed, as if the whole ocean were draining away into some hole. Fish, trapped by the waters' disappearance, flopped helplessly on the sea floor. Delighted by this bonanza, groups of men rushed out to gather the stranded fish. But others had a presentiment of what the sea's retreat meant: a tsunami was on its way.

Acting on their fears, some began to run inland. One villager climbed the slope above Kahului and, hearing the roar of the ocean behind him, turned, and saw an unbelievable sight. People, houses, canoes and animals were riding up the slope on a surge of black, angry water. All Kahului was carried 240m/ 800ft inland by the tsunami.

The popular picture of a tsunami is a scaled-up version of the familiar waves that pound against coasts the world over. It is imagined as a giant wall of water, rushing inland, towering over trees and buildings, then thundering down on them like a gigantic breaker. As the experience of the Kahului villagers illustrates, most tsunamis are not, however, advancing walls of water, but a sudden upwelling of the ocean.

The upsurge spent, the ocean retreats as rapidly as it came, often with a terrible sucking noise as it claims its victims. The suction is so powerful that no swimmer can resist its pull. But this is never the end of the disaster. The popular idea of the tsunami as a single apocalyptic wave is also false. Tsunamis come in trains of a dozen or more waves, and there is no telling which will be the strongest. There may be as little as five minutes between each wave, or as much as an hour. Occasionally tsunamis do produce a ''wall of water'' effect,

The Japanese have intimate knowledge of the tsunamis, or ''harbour waves'', that afflict their country. The artist Hokusai's woodblock print, The Wave, *brings home with wrenching intensity the plight of men at the mercy of such a giant wave. However frantically they paddle, they cannot escape its implacable force as it sweeps shoreward.*

similar to a tidal bore. But this only happens if the wave hits a narrow inlet at a particular angle.

One of the most curious aspects of tsunamis is that they usually pass under ships in mid-ocean without any noticeable effect, and are indistinguishable from ordinary waves. Sailors anchored well offshore have reported watching a giant wave devastate the coastline, while feeling nothing themselves, though directly in the tsunami's path.

Again, it is the topography of seabed and shore that causes the trouble. Where the tsunami collides with small islands or certain types of coastline, it dissipates its energy with minimal effect. But where there are bays and inlets, and a gradually sloping sea floor, the tsunami's prodigious force is channelled into a small space, and confined by the coastline until it reaches breaking point.

In some ill-fated spots, two different parts of the coastline reflect the tsunami waves so that they meet at some third point with amplified force. Like a magnifying glass focusing the Sun's rays, so the coastline focuses the tsunami on one spot, where the water rises to swamp the shore with frenzied violence.

The secret of the tsunami's terrible power lies in its depth. Other waves are superficial murmurs on the water's surface, whipped up by wind and storm. The tsunami pulses through the ocean from its floor to its sunlit upper waters. The effect on the surface is small—a slight, unremarkable, rise and fall as the tsunami wave train passes by. The rise is only about 20–30cm/8–12in, and although this wave train travels as fast as a jet aircraft, at 700kmh/440mph, its progress goes unnoticed in the mêlée of ocean waves. Only when it meets a gradually shelving shore does the power of the tsunami reveal itself.

Most tsunamis have their origins in the Earth's core, where movements of the magma force the plates of the Earth's crust to jolt and jar against one another. Those movements may produce a rise in the seabed as part of the crust is forced upwards, or a sinking of the seabed as one continental plate succumbs to the pressure of another.

The seabed acts like the skin of a drum when struck forcefully by a drumstick. Just as the movement of the drumskin sets up reverberations in the air around it (which we hear as a sound), so the movement of the seabed sets up vibrations in the water above. A tsunami has begun—and an earthquake, both powered by the same crustal movement.

An earthquake vents its force near its point of origin, but a tsunami can travel for thousands of miles across the ocean, to arrive without warning on some distant shore. Hawaii suffers through being at the very centre of the Pacific's "Ring of Fire", a target for tsunamis from every point of the compass.

But not all tsunamis are such long-distance effects. Many earthquakes in coastal areas are followed soon after by a devastating tsunami generated by the same movement of the seabed. Indeed, the Alaskan quake of Good Friday, 1964, was followed by a tsunami that engulfed buildings three floors high.

In Seward, the earthquake shattered dockside pipes leading to oil storage tanks, and the oil caught fire, causing explosions that spewed burning oil across the water. The next wave carried with it a layer of burning oil, "like a huge tide of fire washing ashore."

Subsequent tsunamis spread the destruction far and wide. The steel tracks of the railroad glowed crimson in the blaze, until they were inundated by a wave, when the sudden cooling of the water made them "hiss and curl like snakes" rising from the ground.

Lisbon, the seaport capital of Portugal, was ravaged by a double tragedy on 1 November 1755. A submarine earthquake shattered the city, and as the terrified people fled into the streets, a giant 15-m/ 50-ft tsunami surged ashore, followed by two more. The wave train smashed ships, washed them ashore, then retreated, sucking up debris and people; and fire broke out. In all, more than 60,000 people lost their lives.

A giant wave, welling up in the shallow waters of the Sunda Strait, is proof indeed that a tsunami can be a veritable "wall of water". In the background looms the volcano Krakatoa.

A volcanic eruption under the sea, left, displaces water over the site, which dissipates as a succession of huge waves.

Earthquakes in the sea floor have a similar effect, below. The shift in the seabed creates a bulge in the water. The resulting waves are harmless unless they reach shallow coastal waters, when tsunamis may form.

Seamounts and submarine canyons

Human beings are land creatures, and the processes that go on above sea level tend to have more significance for us than hidden, underwater ones. Our imagination is immediately gripped when we learn that, due to the meshing of movements and forces known as plate tectonics, Europe and North America are edging apart at the rate of around 2cm/1in a year—at the rate our fingernails grow, as one scientist has put it. Much more difficult to grasp is the concept that the mighty forces capable of pushing continents across the face of the Earth are largely created, unseen and rarely revealing themselves, under the great oceans.

When we look at the surface of the sea, all is motion. Waves, currents, swell, tides and spray combine to give an abiding impression of ceaseless watery activity. And until about 30 years ago, people believed that the state of the sea floor was the complete opposite of that of the waters above it. The bedrocks of the ocean were thought to be ancient and immobile foundations upon which never-ending drifts of sediment particles settled. But the science of plate tectonics has revealed that nothing could be farther from the truth. The ocean floors are all young rocks, none more than around 200 million years old. They are also all in motion.

Most Earth scientists believe it is the movement of the ocean floor that provides the main motive power for all the other crustal jostlings making up the complex patterns of continental drift, or plate tectonics. The rock layer concerned is termed the "lithosphere", a stiff shell about 75km/47mls thick, which is made up of the crust itself and the outer layer of the mantle beneath it. Rising up 3km/2mls from the crust at the bottom of the ocean basin is an almost completely unseen series of underwater mountain ranges. In broken

Fired by a "hot spot" that acts like a gas jet on magma deep within the Earth, the volcanoes of Hawaii spew out red-hot lava in an almost uninterrupted stream. Billowing clouds of steam rise where the lava flows into the sea and cools. This continuous process of island-building provides an uncanny and awe-inspiring insight into the primeval past of our world, for in just such a way the first dry land was created.

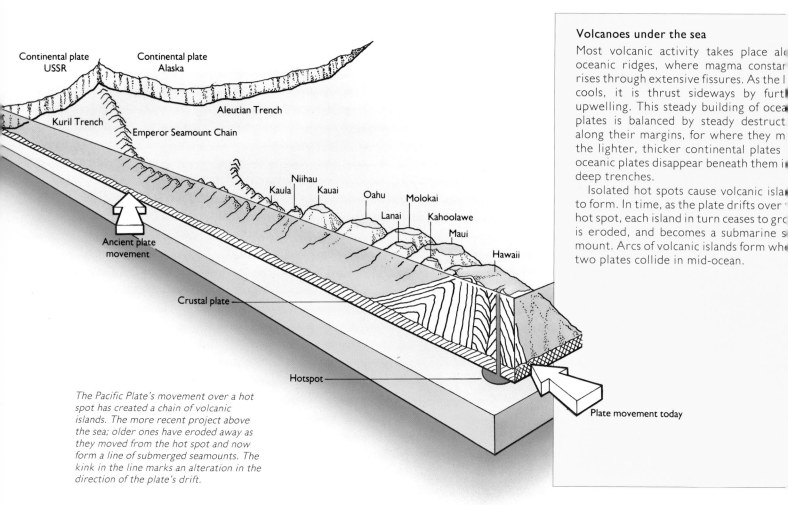

Continental plate
USSR

Continental plate
Alaska

Kuril Trench

Aleutian Trench

Emperor Seamount Chain

Kaula
Niihau
Kauai
Oahu
Molokai
Lanai
Kahoolawe
Maui
Hawaii

Ancient plate
movement

Crustal plate

Hotspot

Plate movement today

The Pacific Plate's movement over a hot spot has created a chain of volcanic islands. The more recent project above the sea; older ones have eroded away as they moved from the hot spot and now form a line of submerged seamounts. The kink in the line marks an alteration in the direction of the plate's drift.

Volcanoes under the sea

Most volcanic activity takes place al(oceanic ridges, where magma constar rises through extensive fissures. As the I cools, it is thrust sideways by furtl upwelling. This steady building of ocea plates is balanced by steady destruct along their margins, for where they m the lighter, thicker continental plates oceanic plates disappear beneath them i deep trenches.

Isolated hot spots cause volcanic isla(to form. In time, as the plate drifts over hot spot, each island in turn ceases to grc is eroded, and becomes a submarine s mount. Arcs of volcanic islands form wh(two plates collide in mid-ocean.

chains, they encircle the planet more than once. Known as ridges, or rises, these ranges are crucial zones in the landscape of plate tectonics.

The mountain ranges tend to lie in mid-oceanic positions. In the North Atlantic they are known as the Reykjanes Ridge, in the South Atlantic as the Mid-Atlantic Ridge, and in the Pacific as the East Pacific Rise. At every ridge, along these narrow strips of the Earth's surface, molten rock from the mantle rises, cools and produces new lithosphere. For this reason, such lines of activity are known as the constructive margins of plates.

This constructive activity builds the undersea mountain ranges. Vast seethings of molten magma are ceaselessly solidifying into submarine mountains all around the world. At the ridges, the magma cools, solidifies and then moves symmetrically sideways to form each flank of the ridge. Behind it rises more solidifying rock, pushing the band ahead

of it farther sideways and producing new ocean floor crust. Such spreading at the constructive margins in the middle of the North Atlantic is forcing New York and London slowly apart.

Slow spreading enables the lagging rocks of the margin zone to pile up into a mountain ridge. Faster spreading moves the new rock more speedily sideways, so the mound at the margin is lower and forms a so-called rise, as in the East Pacific Ocean.

Most of this mysterious and potent activity is forever hidden from human eyes. In a few places, however, a midoceanic ridge emerges from the sea as a volcanic landscape. Two such bizarre locations are the islands of Tristan da Cunha and Iceland, both of which sit astride the mid-Atlantic constructive zone and are still being built by it.

Iceland is a young island; it has been constructed out of the cooled magma from a 322km/200ml long emergent portion of the Reykjanes Ridge. The

country is neatly bisected by an oblique line of active volcanoes, running southwest to northeast, which mark the plate margins. On either side of this line is a strip of countryside composed of recent Quaternary rocks—beyond them the margins of the islands are older Tertiary rocks. On the sea bottom we may suppose that there is a similar congruence between the distance from the ridge and the age of the rocks: those closest are youngest, those farthest away oldest.

A study of paleomagnetism, the magnetism remaining in the rocks from past times, confirms this expectation. As the molten ridge rock cools, it retains an internal imprint of the direction of the magnetic poles of the Earth at the moment of its cooling. When the magnetic poles switch position every few hundred thousand years, the magnetic imprints in the rocks cooling at the time also switch.

Once it became possible to map these changes in rock samples brought up from

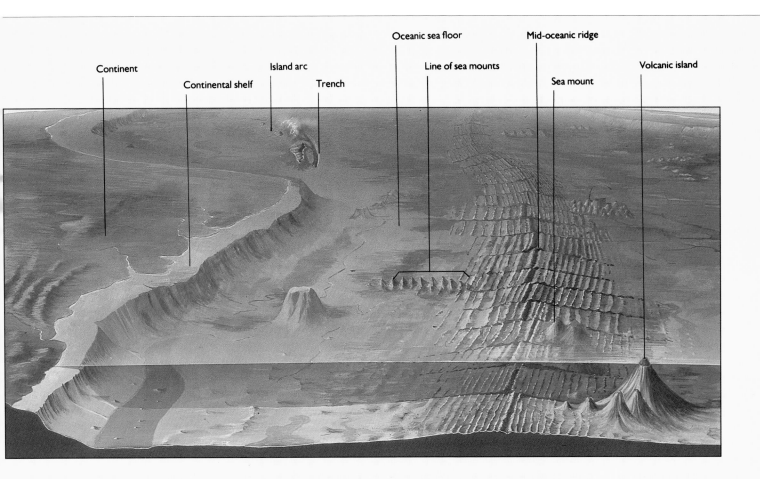

Continent   Continental shelf   Island arc   Trench   Oceanic sea floor   Line of sea mounts   Mid-oceanic ridge   Sea mount   Volcanic island

beneath the ocean, it became clear that such switches in magnetism extend for hundreds of miles on either side of mid-oceanic ridges. More recently, sensitive instruments, towed behind ships, have been used to map the magnetic reversals in the seabed rocks along the ship's course. The covert patternings in the rocks that these oceanographic cruises have discovered can be explained only if the ridges have been centers of rock spreading for many millions of years.

Further evidence of the constructive powers of magma lies off the southern coast of Iceland. Here, in 1963, a new volcanic island, Surtsey, emerged from the sea, an event that dramatically demonstrates how magma flows near constructive margins can produce new underwater mountains. Similar volcanic mountain construction can occur over so-called "hot spots" beneath the previously solidified sea floor crust. One of the best examples of such a hot spot is at present situated under Hawaii.

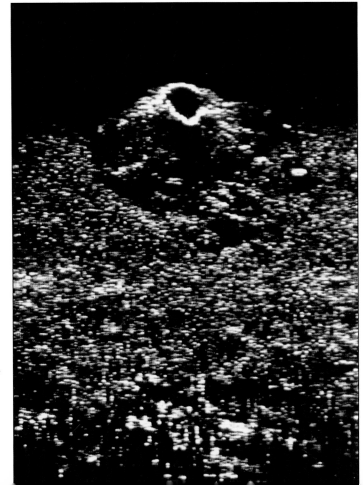

As astonishing as the great underwater mountains themselves is the scientists' ability to gather information about them. The sonagraph, left, made by long-range side-scan sonar, shows an undersea volcano on the East Pacific Rise with a crater 2km/1.2mls wide and, possibly, a moat. Lying under some 4,000m/ 12,210ft of water, the volcano stands 1,500m/4,921ft high and measures 10km/ 6mls across at the base.

It seems that although this hot spot itself is fixed, its position relative to the Earth's crust has been changing for tens of millions of years. The huge Pacific crustal plate has, however, been moving north and west over it in the approximate direction of Japan. This has led to the formation of the Hawaiian Island Chain, which extends in a band from Hawaii west to Midway and Kure islands, some 2,250km/1,430mls away.

Still farther west and north, but more or less continuing the line, is the so-called Emperor Seamount Chain, none of whose mountains breaks the surface. If the ages of the rocks making up all these islands and seamounts are studied, a sequence becomes clear. Hawaii itself is young, with all parts less than 700,000 years old and some regions still being formed by volcanic action. Proceeding north and west, the rocks become steadily older, until the Meiji seamount, with rocks scientifically dated as 70 million years old, is reached.

What has caused this mysteriously straight line of islands and why are their ages so neatly graded? The answer is the hot spot. The Pacific Plate has moved over it fairly steadily, at some 8–10cm/3–4in a year. Every million years or so, sufficient lava is produced over the hot spot to create an emergent volcanic island, in just the way that Hawaii is still forming.

But as the plate moves, it carries the island along with it and eventually severs it from its source of new, molten rock. The island then stops growing and the forces of rain, wind and ocean waves start to wear it down. As the island is rafted northwest, it gets lower and lower in the water until, after some 30 million years on average, it sinks beneath the waves to become a seamount.

This presumed interaction of hot spot, plate motion and erosion could explain both the linear neatness of the island chain and its age sequence. If the

suggestion is correct, one could imagine that a "son of Hawaii" would eventually form; and, just as the theory predicts, there is indeed a 300m/985ft high mound on the seabed southeast of the present island. The hot spot has started to produce another island.

The recently discovered, intense volcanic activity of the oceanic constructive plate margins has been the focus of frenetic geophysical investigation. Among the most spectacular studies have been those conducted using a series of new-technology, deep-sea submersibles. These investigations have demonstrated that strong heat flows occur in the rocks at the ridge. Careful descriptions have been made of the different forms of rapidly cooling lava found when molten rock emerges on the seabed. Among the most unusual are the pillow lavas—rounded chunks of lava that cool separately, then lie together in mounded heaps.

The most exciting discovery made so

Explorer, ornithologist and writer, William Beebe (1877–1952) was one of the first men to descend deep into the ocean.

Sea Link's plastic dome gives a superb all-round view. A rear pressure chamber enables divers to work from the craft.

Exploring the ocean depths

Underwater exploration came into its own in 1934, when the Americans William Beebe and Otis Barton were lowered by cable to a depth of 923m/3,028ft in Barton's heavy steel bathysphere. Twenty years later, in a maneuvrable bathyscaphe, the Belgian Auguste Piccard and his son Jacques reached a depth of 3,167m/10,392ft off the Italian coast. In 1960 Jacques plumbed the extreme ocean depth of 10,912m/35,800ft in the Marianas Trench.

far by the submersibles has, however, been the deep-sea vents. In highly volcanically active sea floor zones, the underwater craft have found large numbers of individual jets of hot water emerging from the rocks into the surrounding ice-cold depths. These vents seem to be due to seawater entering clefts in the hot rock, where it is heated to fairly high temperatures and subsequently expelled through the vents.

In its passage through the hot volcanic rocks, the water picks up large concentrations of dissolved and suspended minerals, particularly iron, zinc, manganese, and copper sulphides. These turn the water into a dense cloudy suspension which, when it is forced out of the vents, emerges as white or black underwater "smoke". The jets of water from the vents are often called "smokers" and sometimes the minerals in the stream deposit out as solids around the vent to form hollow columns, or chimneys. These amazing structures and flows

would be remarkable even if they were purely geophysical phenomena, but their presence has engendered communities of underwater life that are among the most unorthodox on our planet.

Everywhere else on the globe, communities of organisms are, in the final analysis, dependent for energy on the Sun. On land, in marine or freshwater, it is normally green plant life that, by trapping sunlight by photosynthesis, provides the raw material to sustain all other organisms in the community. Herbivorous animals eat the plants and are, in turn, eaten by carnivores.

The vent communities are quite different. In the total darkness of the ocean depths, high concentrations of sulphides in the hot-water jets appear to provide an alternative source of energy for some chemical-using bacteria. Such bacteria, which can split hydrogen sulphide to gain energy, appear to be the living base of the vents' animal communities. They are fed on by a variety of filter-feeding shellfish found only at the vents. In some instances the bacteria have entered into a symbiotic relationship with the vent animals, living on their tissues and, in turn, providing

them with the nutrients they require.

This is apparently what happens with the bizarre pogonophorans. These giant, bright-red tube worms, some $2m/6\frac{1}{2}$ft long, have no gut, yet they grow and cluster thickly around some vents. The secret of their vitality seems to be the sulphide-busting bacteria in their bodies. Thousands of feet below the waves, in utter blackness, crimson worms live out a life based on hot, sulphurous mineral water formed by percolation through volcanic rock. This must surely be one of the Earth's strangest and most incredible mysteries.

Scientists exploring the normally barren sunless depths of the Galapagos hydrothermal region in 1979 came upon the startling sight of giant red tube worms clustered around deep-sea vents. The mystery increased when it was discovered that the worms had no mouth, no gut and no anus. Rather than feeding in a conventional way, they survive in a symbiotic relationship with bacteria in their body cavity, which convert sulphides in the water into nutrients. Crabs and mussels, below, may subsist in a similar fashion.

Into the icy waters around the East Pacific Rise, ocean "smokers" eject black plumes of mineral-rich water, heated in some places to 350°C/662°F. Pressures are so high at such depths—some 2,600m/8,530ft—that the seawater does not boil.

Hot plume of heated seawater carrying dissolved minerals.

Deposited mineral chimney

Superheated seawater exits through vent

Cold seawater enters rock fissures

Hot molten rock

Deep-sea "smokers"

This highly diagrammatic view of a vent chimney, deep under the ocean, shows how almost freezing water (around 2°C/35.6°F) percolates into porous rock on the sea floor. It flows through the rocks, where it is heated by magma rising along the spreading ridge, and then emerges at a vent. The hot, turbid seawater is rich in minerals, especially sulphides, which it has dissolved from the rocks, and these precipitate out to form the chimney wall. Such conditions have given rise to the extraordinary community of life forms that flourishes around the hot, deep-sea vents.

The icebound continent

Antarctica, the continent of ice, is separated from Tierra del Fuego, the land of fire, by a channel of turbulent, stormy sea no more than 1,000km/ 620mls wide. Here, at the tip of South America, is the point where Antarctica comes closest to the rest of the world. It was across this channel that the first ships ventured, or were driven by the winds, to come up against the great surrounding landmass the South Pole.

Nineteenth-century whalers and sealers, those first explorers were tough adventurers in search of new hunting grounds as they depleted the seas and islands farther north. They had no idea that a continent lay in wait for them. And in the shifting, deceptive seascape of icebergs, blizzards and fog, without any map or chart, these men could make little sense of the mysterious land they glimpsed in the distance.

Earlier, with their passion for logic and symmetry, the ancient Greeks had concluded that there must be a vast southern continent to balance the northern world of Eurasia. This idea remained largely unchallenged until 1578, when Sir Francis Drake was pushed far south by violent storms, but saw no sign of the mythical Terra Australis.

Mysterious and almost unbelievable in the surrounding wilderness of ice, the 3,794-m/12,444-ft volcano Mount Erebus puffs out a plume of smoke. In 1841, when the explorer Sir James Clark Ross first saw Erebus—one of four volcanoes on Ross Island—flame and smoke were shooting hundreds of feet into the air. It last erupted in 1984/5, with violent explosions of gas.

Strange and beautiful optical phenomena, such as the bright spot on the solar halo, or parhelion, below, are often visible in the clear, dry Antarctic atmosphere. The parhelion is created by rays of the Sun, here low in the sky at 2am on an Antarctic summer night, shining through ice crystals in the air.

Two centuries later, in 1778, Captain Cook sailed south and spent three futile years in search of the southern continent, crossing the Antarctic Circle three times. In the end, he was forced to admit that Terra Australis did not exist. But what Cook had seen in the southern seas convinced him that there was something there—"a Tract of Land near the Pole, which is the Source of all the Ice spread over this vast Southern Ocean." It is only in the twentieth century that we have come to appreciate how prescient Cook's words were. Antarctica is indeed vast, larger than Australia or Europe, larger than the United States and Mexico combined.

Ice forms a layer over the Antarctic landmass that is, on average, 2,400m/ 8,000ft thick, although over one deeply gouged valley it measures almost 4,800m/15,800ft. The pressure at the base of this pile, where the ice bears down on the rocks below, is immense. Indeed, scientific measurements and satellite observations have shown that the Antarctic is forced downward into the Earth by its burden of ice, giving our planet a flat-bottomed, pear-shaped outline. If the ice were suddenly to melt, Antarctica would spring upward by as much as 600m/2,000ft.

All this ice fell as snow, some of it

An iceberg, "calved" from the Ross Ice Shelf, floats freely in the sea, above. Depressed by the great weight of ice, the bedrock of Antarctica lies well below sea level, although the average altitude is greater than that of any other landmass.

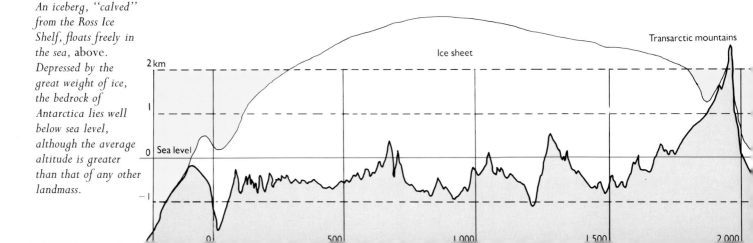

millions of years ago. Compacted by new layers pressing down from above, the snow gradually loses its air pockets and consolidates into ice. As the ice builds up, its own weight forces it down from the central heights to form glaciers. Some unload their frozen cargoes directly into the sea, but others form massive ice shelves, floating rafts of ice many thousands of miles across.

Glaciers move with unimaginable slowness—sometimes less than 1m/3ft a year—as the ice inside them becomes plastic under intense pressure. Unlike rivers, they have no steady input of water at their head, for snow in Antarctica falls on the coast, and the heart of the continent receives hardly any moisture. It is drier than the Sahara Desert, and only the extraordinary slowness with which the glaciers move keeps the land covered in ice. In this strange corner of the world, the extreme cold seems to have frozen time itself.

As well as being the driest place in the world, Antarctica is also the windiest, with gusts of up to 320kmh/200mph. If they carry snow, these turn into blinding "white winds", in which it is impossible to navigate even the shortest distances. The winds can also have a bizarre action on the snow, sculpting its upper layers into miniature ridges and peaks.

No place on Earth is as unwelcoming to life as the Antarctic landmass. Apart from drifts of resilient red-coloured algae, which can survive on the surface of the snow and ice, life is confined to a few spots that are ice-free in summer.

These oases in the frozen desert are found mainly along the coast, especially on the finger of rocky land, known as the Antarctic Peninsula, which reaches out toward South America. These spots have

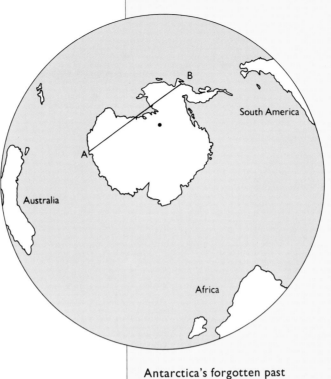

Until 150 million years ago, Antarctica was joined to India, Africa, Australia and South America in the supercontinent of Gondwana. As Gondwana broke up, continental drift forced Antarctica south, and eventually it settled over the South Pole. The cross-section below represents the line A-B.

Antarctica's forgotten past

In 1969, high on the Transarctic Mountains, fossilized bones were discovered. These were not fossils of penguins, seals or other polar creatures, but of *Lystrosaurus*, an animal that lived 220 million years ago. Neither a reptile nor a mammal, but somewhere between the two, *Lystrosaurus* stood 50cm/20in tall and probably lived a semi-aquatic existence in pools of warm water. This, and other fossils of ferns, crocodiles, turtles and trees show that Antarctica was once subtropical and was, in fact, part of Gondwana, for they are similar to fossils found in the southern continents. When Gondwana fragmented, Antarctica drifted slowly southward, became cooler and, for a time, enjoyed a temperate climate. Finally it settled over the South Pole and the ice cap began to form. As it did so, this ice cap cooled the planet, like a giant refrigerator. It caused tropical forests to give way to grassland in large parts of Africa, probably providing the impetus for tree-dwelling apes to evolve into the plains-dwelling ancestors of "Modern Man".

Artist's impression of Lystrosaurus

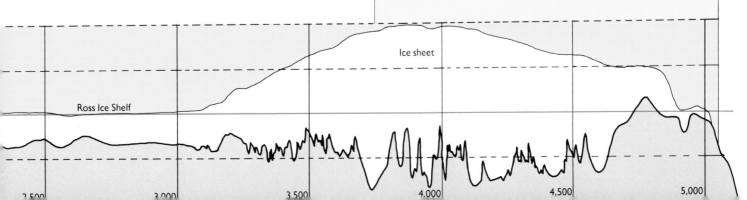

Ross Ice Shelf

Ice sheet

2,500 3,000 3,500 4,000 4,500 5,000

Norwegian whalers made the first landings on the Antarctic mainland in 1895. Interest in the continent grew rapidly, and in 1901, a British scientific expedition, led by Captain Robert Scott (1868–1912), set sail for Antarctica in the wooden ice-breaker Discovery, right. *They spent more than two years exploring and researching.*

Scott's second expedition of 1910 coincided with that of the Norwegians, under Roald Amundsen (1872–1928), left. The race to the South Pole was on. Using dogs, Amundsen reached the pole on 14 December 1911, Scott on 18 January 1912. Beaten in the race, and finally defeated by blizzards and temperatures as low as −40°, Scott and his four companions died on their journey back to base.

the most luxuriant vegetation, for a species of grass, Antarctic hair grass, *Deschampsia antarctica*, survives here, together with dense hummocks of moss which add a rare touch of green to the landscape. Elsewhere, there is nothing but lichens—composite life forms based on a tough, enduring alliance of fungi and algae. Only lichens can survive the harsh, arid conditions.

Animal life is there on a Lilliputian scale. Simple roundworms, mites, and primitive insects, known as springtails, are the main inhabitants of the soil and vegetation. Some can be seen with the naked eye; but even the largest, *Belgica antarctica*, a flightless midge, measures no more than 13mm/½in. Most are visible only with a microscope.

Other pockets of life are found far inland, some quite close to the South Pole, where the highest peaks of the

Transantarctic Mountains breach the blanket of ice. These peaks are known as *nunataks*, a name borrowed from the Inuit language by some early explorers. There are no large plants found here, only microscopic algae and fungi, bacteria and some minute animals.

Least promising of all the oases are the Dry Valleys, found only in a small area of Victoria Land, due south of New Zealand. There are just three main valleys, gouged out by glaciers long ago. Only a thin dusting of snow ever falls here today, and it has been thus for at least two million years. In that time, the unreplenished glaciers have vanished, and what remains is a bleak landscape, bitterly cold and completely arid, where ruthless winds scour the rocks.

Until 1978, the Dry Valleys were thought to be utterly devoid of life. But then scientists examined the rocks with

powerful microscopes and found one of the most unlikely living communities on Earth. Beneath the rock surface, in minute pockets of air, there were bacteria, fungi and algae, continuing a frugal way of life that must have begun millions of years ago.

Trapped within the rocks, these microscopic survivors depend on carbon dioxide gas seeping through to them, and on sunlight, filtered by the rock to a dim glow. This glimmer of light allows the algae to make small quantities of food by photosynthesis, and the bacteria and fungi tap into this food source. Like all living things, these organisms require water. Tiny beads of moisture, derived from infrequent snowfalls, penetrate the rocks, and the cloak of rock prevents the creatures' fragile bodies drying out.

Life on the land may be sparse, but the waters of the Antarctic are among the

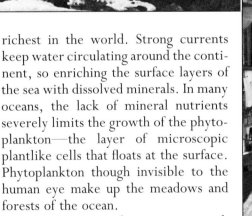

richest in the world. Strong currents keep water circulating around the continent, so enriching the surface layers of the sea with dissolved minerals. In many oceans, the lack of mineral nutrients severely limits the growth of the phytoplankton—the layer of microscopic plantlike cells that floats at the surface. Phytoplankton though invisible to the human eye make up the meadows and forests of the ocean.

Temperatures in the ocean are much higher than on land, and there is uninterrupted daylight to fuel photosynthesis during the summer months. The abundant phytoplankton fall prey to the small animals of the zooplankton, including vast shoals of krill. These small, shrimplike creatures flourish in the Southern Ocean, congregating in shoals that may cover 2,500sq km/965sq mls. They give a pinkish tinge to the

Scott, seen here in the hut at Cape Evans, kept his diary assiduously; his last entry, on 29 March 1912, reads: "we are growing weaker . . . the end cannot be far."

water during the day, and light up the sea like aquatic fireflies at night.

Krill are the bread-and-butter of Antarctic waters, a single source of food for animals of all sizes, from gigantic blue whales down to pint-sized Adelie penguins. Five species of whale and four species of penguin rely on krill, along with innumerable other birds, three species of seal, and at least twenty species of fish. There is nowhere else on Earth where a single food source supplies so many different creatures, and the reasons for this unusual situation remain a mystery.

There is little doubt that the heavy reliance on krill makes Antarctic wildlife uncommonly susceptible to disruption of the food chain by humans. To some, the netting of krill for human food may be attractive, but to biologists it has the makings of an ecological disaster.

Antarctica is largely contained by the Antarctic Circle, the point at which the Sun appears to turn back on its yearly journey across the sky, where it remains rooted on the horizon on midwinter's day. Nearer the pole, winter brings with it a period of days or weeks when the Sun fails to rise, and the only light is the eerie glow of the aurora australis.

Antarctica is still the ultimate adventure for many explorers, scientists and even tourists, and exploration is well organized and highly mechanized. But increasing numbers of visitors and exploitation of this magnificent icy land, unless strictly controlled, will despoil it and destroy its fragile ecosystems. In the cold dry air, the processes of decay are arrested. Thus the debris of the abandoned whaling station, opposite, will remain for years as a monument to human desecration.

The Antarctic Circle is, however, only one of the many circles that contain and define this continent. The South Pole marks the axis on which the Earth spins; and, in their own fashion, the air and ocean also rotate about that axis, creating circular weather systems which dominate Antarctica. Unhindered by any great landmasses, such as those found in the Northern Hemisphere, the circumpolar winds whistle around the globe, their fury unabated.

Another unwelcoming circle, known as the storm belt, is created by cold

Antarctic scientific research is wide-ranging. It extends under the sea, with divers charting the depth of the ice sheet and monitoring the stocks of fish. And on land, mineral research has turned up more than 7,500 nickel-iron meteorite fragments which, since they are only 1,300 million years old, younger than any others yet found, are probably from the Moon or Mars, which was still volcanically active at that time.

winds flowing down from the heights of the Antarctic under the force of gravity. Where they meet the relatively warm air above the ocean, the clash of atmospheric personalities generates dense fogs, cloud and terrifying blizzards.

The ocean, likewise, swirls in great circles about the continent, though its movements are more complex. Close to the coast, currents flow from east to west, taking with them icebergs that have calved from the glaciers. But about 2,000km/1,200mls from the coast, this cold water meets warmer water circulating from west to east. Being denser, the cold water sinks below the surface, and icebergs which have floated this far northward abruptly change direction.

North of this meeting point, which is known as the Antarctic Convergence, the water holds different minerals and supports a different mix of life forms. The huge shoals of krill are found only within the convergence, and the animals that feed on krill are similarly confined there during the brief summer. This invisible meeting point in the oceans, which is just a few miles wide, surrounds Antarctica like a city wall, marking the approaches to the Earth's least known, ice-bound continent.

The Antarctic Treaty of June 1961 established the international status of the continent. The conflicting territorial claims of the nations involved were recognized and defined; military operations, nuclear explosions and the dumping of nuclear waste were banned; and freedom of movement for scientists, as well as the exchange of data, was introduced.

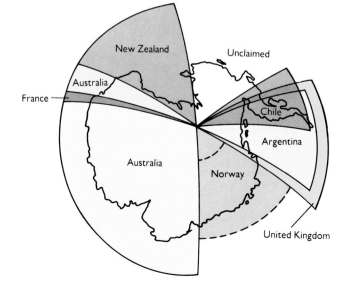

The weather machine

The weather affects all living things on our planet. At its extremes, the weather can bring flood or drought, or wreak havoc through the incredible power of tornado and hurricane. Weather helps determine the landscapes and economies of nations and, on a more parochial daily level, can cause traffic chaos, make or break a vacation, or even alter our moods. No wonder, therefore, that we remain fascinated by its influence and its unpredictability.

Humans have long been intrigued by the mysteries of the weather. Since ancient times, gods have been associated with its powerful and recurring wonders—thunder and lightning, frost and snow, wind and rain. And people everywhere have felt the need to propitiate the gods, whose might determines the difference between famine and plenty and whose anger can spark the sky.

With practical acumen, our ancestors could, by careful observation and the benefit of experience, discern definite patterns in the weather, from which they would make remarkably well-informed judgments about such matters as the best planting or harvest times. Today, even with the benefit of the satellites, tracking instruments and other technological wizardry of modern meteorology, much of the ancients' weather lore still holds good. Equally, many aspects of the weather remain unexplained.

Weather is not something restricted to the Earth—any planet that spins and has an atmosphere has weather. Jupiter, Mars, Saturn, Venus and Neptune, which have inhospitable and utterly different weather worlds from our own, prove that this is true. The general nature of the weather depends on the temperature of the planet, the composition of its atmosphere and the amount of water present at its surface.

The key to Earth's weather lies in the fact that the temperature of the Earth is warm enough for most of its water to be in the form of liquid, in contrast with the supercooled ice present on Mars or the superheated steam on Venus. From space, our planet looks blue, speckled with white— a remarkably beautiful watery world of oceans and clouds.

The weather we experience is generated by a

Swirling, wind-blown rainclouds characterize planet Earth.

surprisingly thin layer of atmosphere, which extends only 18–30 km/11–19 mls above the Earth's surface. The clouds and air circulations of our planetary weather systems are confined within this troposphere. Here, the atmosphere is hardly ever still, for winds blow over almost all parts of the Earth's surface all the time.

Winds, the invisible forces that help shape our weather, are caused by differences in atmospheric pressure. A wind will always blow from a region of relatively high pressure toward an adjacent region where the pressure is lower. But what creates these differences in pressure in the first instance? At its simplest, the reason is that the equatorial zones of the Earth receive more light energy, in the form of heat from the Sun, than do the regions near the poles. Warmed air near the Equator rises, moves poleward, then cools, sinks and returns to its origin.

The real picture is, however, much more complex than this, and is due in part to the Earth's spin. This spin produces the Coriolis force, which deflects the major wind systems into a series of obliquely moving bands of winds around the planet.

Superimposed on the weather zone bands that ring the world is the annual cycle of climatic change, determined by the way that the Earth's spin axis is tilted. It is this tilt which, in the course of 12 months, moves any region through its yearly sequence of climatic seasonal change, with the weather bands becoming crowded together in one hemisphere and expanded in the other as the seasons progress from winter to summer and back again.

The weather is a constant source of wonder to lay observer and meteorologist alike. As we unravel its whys and wherefores and become weather-wise, so we become aware of its delicate balance and of the potential perils of disrupting that balance.

Powerhouse of the winds

Weather would not exist without winds. It is the winds that drive our global weather machine yet, after centuries of study, our understanding of wind patterns and precisely how the winds create the weather remain at least in part an unsolved mystery. Modern technology provides us with breathtaking satellite views of the global atmosphere, but it is the clouds and storms and not the winds that we can see in these images. And because our knowledge of the winds is incomplete, it is not surprising that our weather forecasts contain many uncertainties.

Our own inability to make accurate predictions about the weather make it all the more remarkable that earlier peoples were able to progress so far in fathoming the mysteries of the global weather machine. The ancient Greeks took the first tentative steps by recognizing the importance of winds, which they regarded as superhuman beings.

In 40 BC, the winds were immortalized in a sculptured frieze on the octagonal Tower of the Winds, erected near the Acropolis in Athens. On top of the tower was a weather vane, which pointed to the wind-figure on the appropriate face of the building and so indicated the weather to be expected. Thus Notos (south face) signified rain and Lipos (southwest face) meant good sailing.

The greatest leap in understanding of the world's winds and weather came with the Spanish, Portuguese and English explorations of the New World, beginning in the fifteenth century. Sailing ships would often choose to sail south from Europe, rather than due west, to catch the reliable northeasterly winds, or Trades, which flow from the coast of Africa to the Caribbean.

It took the insight of a brilliant English instrument maker, George Hadley (1682–1744), to explain why the

The weather is the ultimate unknown, shaped by the invisible. For it is the winds, from gentle breeze to roaring blast, that fashion the Earth's weather. Unseen, except in their passing effect on the world around us, they circulate masses of warm air from the Equator and cool air from the poles; they bring life-giving rain clouds and drive devastating storms. Without the winds, the mystery of the weather would not exist.

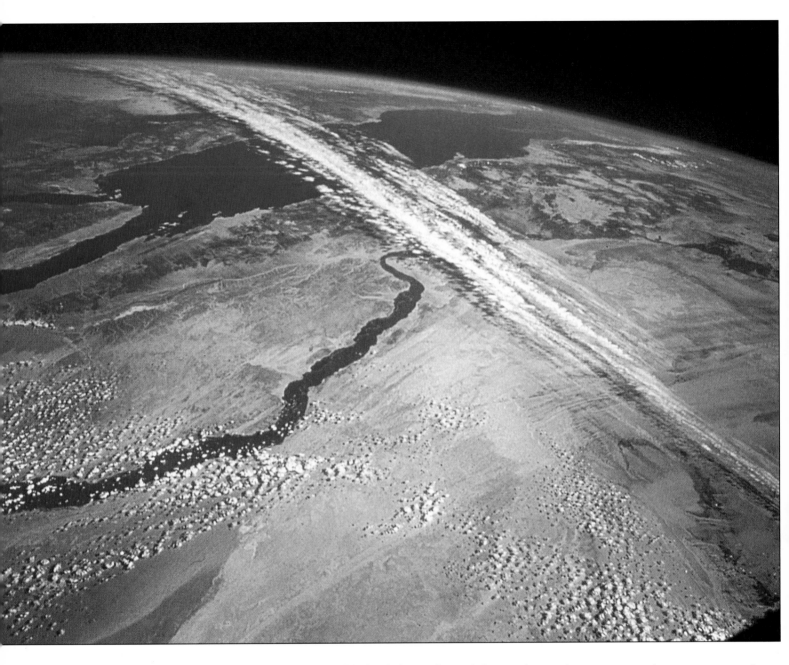

Trade winds existed, and why those in the Northern Hemisphere blew from the northeast and in the Southern Hemisphere from the southeast. In 1735 he suggested that intense solar heating of the equatorial regions causes warm air to rise and flow toward both poles. As it cools, the air sinks and is drawn back toward the Equator, replacing rising equatorial air, to form the Trade winds.

Hadley deduced that the Trades of both hemispheres blew toward the Equator from an easterly direction because the Earth rotates toward the east. He explained that air from higher latitudes blowing toward the Equator lagged behind the surface of the Earth, which rotates more rapidly at the Equator than in higher latitudes, and so appeared to arrive from an easterly direction. This result of rotation, known as the Coriolis Effect for the French mathematician Gaspard Gustave Coriolis (1792–1843), appears to cause all winds to move clockwise in the Northern Hemisphere and counterclockwise in the Southern Hemisphere.

During the nineteenth century, scientists began to amass vast amounts of information on surface air pressure. Air-pressure measurements indicate the vertical motion of the winds. If pressure is low, it means the air is rising; if it is high, the air is sinking. In 1855, such information enabled an American schoolteacher, William Ferrel (1817–91), to offer the first three-dimensional model of global wind patterns. Like Hadley, Ferrel suggested that warm, moist air rises above the Equator and moves in the direction of the poles. At about 30 degrees latitude in both hemispheres, the air cools and sinks, creating a surface high-pressure zone, from which the winds return to the Equator as the Trades.

These belts of sinking air mark a band of deserts around the world, including

Broad ribbons of cirrus cloud reveal the presence of the jet stream 11–13km/ 7–8mls above the Earth. Generated by turbulent winds, these bands of cloud may stretch for as much as 1,610km/1,000mls.

Disturbing airs

Some winds, particularly hot, dry ones, have such an impact on people's lives that they have been given special names.

The Chinook, called "snow eater" by Canadian Indians, is a warm, dry wind blowing down on the eastern slopes of the Rocky Mountains. Cattlemen welcome it because it can melt a snow cover in just a few hours, uncovering the grass beneath.

The Föhn, encountered in alpine areas of Europe, causes irritability, migraine, insomnia and anxiety in about a quarter of the people. In Switzerland and West Germany, suicides, traffic and industrial accidents all increase when this "witch's wind" blows.

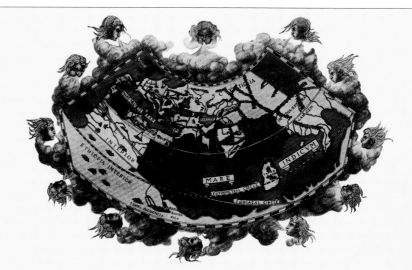

In spring, the Sirocco, or Scirocco, laden with irritating sand, is often pulled north out of the Sahara Desert when a storm rushes eastward through the Mediterranean. As it crosses the sea it absorbs moisture, while still carrying its huge dusty load. Occasionally the iron-stained Sahara dust is washed out as startling red rains over Spain and Italy.

The winds were still envisaged as people when this medieval copy of Ptolemy's second-century world map was made.

pression (low pressure)

yclone (high pressure)

Polar cell

Polar winds

Polar Front Jet Stream

Ferrel Cell

Subtropical jet stream

Hadley Cell

Northeast Trade winds

The Polar Front Jet Stream controls much of the weather in middle and higher latitudes by triggering the formation of frontal depressions and steering the course they take.

the Sahara Desert in Africa, the Mojave Desert of North America and the Great Victorian Desert of Australia. Such parched landscapes have come into being because the air descending from the thin upper atmosphere to the dense lower atmosphere is compressed, warmed and squeezed dry of moisture. In the North Atlantic, this zone became known as the Horse Latitudes, reportedly because livestock being transported from Europe to the New World often died from hunger and thirst as sailing ships became

becalmed for very long periods of time.

Some of the air descending around the 30 degree latitude moves toward the poles, rather than the Equator, and forms the warm southwesterlies and westerlies of the latitudes between 30 and 60 degrees—the middle latitudes. Around the 60-degree latitude, these warm winds, which gather moisture as they pass over the oceans, clash with cold dry air flowing from the intensely cold poles. Ferrel believed that the dense polar air undercuts the warm

westerlies and pushes them up and toward the Equator in a return airflow, so completing a mid-latitude circulation, or "Ferrel Cell".

Ferrel's model of the global weather machine was undisputed for 50–60 years, until meteorologists started to speculate on just how the frequent swirling storms of the middle and higher latitudes contributed to the global weather machine. The answers came during World War I from a Norwegian research group headed by Vilhelm and Jakob Bjerknes.

Like the opposing armies in the war, the mid-latitude westerlies and the polar

2 *A watery sun and thin altostratus clouds, which continue to deepen and lower, mark the approach of the warm front.*

Cold air

Cold front

6

Tracking a depression

Swirling depressions are like huge cauldrons, within which warm and cold winds clash; the leading edges of these winds, or air masses, are known as warm and cold fronts. The air masses slowly mix over several days, but since they retain their temperature and moisture characteristics for much of that time, a distinctive sequence of clouds, rainfall, winds and temperature is produced.

1 *Thin, streaky cirrus clouds herald a frontal depression; but it is still some hours away.*

easterlies were described as doing battle along a "front" at about latitude 60, which the Norwegians called the "Polar Front". This clash of winds results in the creation of massive eddies, many hundreds and even thousands of miles across, which are swept along from west to east. The storms, with their inflowing and rising winds rotating counterclockwise, are known as cyclones, lows or frontal depressions. Interspersed with these lows are the tranquil highs, or

anticyclones, with sinking air and out-flowing light winds.

Despite the advances made in understanding mid-latitude weather systems, the accuracy of weather forecasting remained poor into the 1930s. One reason was that little was known about when a frontal depression would form and what track it would follow over its 3–4 day life cycle. Then, during World War II, a surprising discovery was made.

In 1944, American aircrews on high-

flying bombing raids on Tokyo were astonished to encounter exceptionally strong high-level winds, which increased their flying speed by as much as 160kmh/100mph when they flew east and decreased it by a similar amount when they flew west into these headwinds. Post-war investigations showed that the winds formed narrow meandering ribbons of air, seldom more than 480km/300mls wide and 3km/2mls thick, which meteorologists call jet

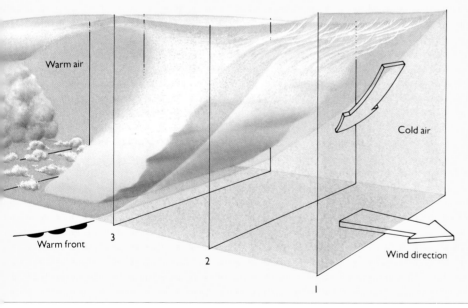

3 *Stratus clouds overhead at the warm front bring several hours of drizzle or moderate rain.*

4 *Shallow stratus and stratocumulus clouds in the warm sector give a brief respite before the rain clouds come again.*

5 *The passing of the cold front is typified by heavy rainfall from threatening black clouds.*

6 *Cumulus clouds, building into isolated cumulonimbus— thunder clouds—give short, heavy showers before clearing away.*

Warm air

Cold air

Warm front

3

2

Wind direction

1

streams. They streak around the globe at speeds of 160–240kmh/100–150mph, even 480kmh/300mph has been recorded—the fastest winds on Earth.

The energy to create the most important of these, the Polar Front Jet Stream, is derived from the steep temperature gradient between the warm westerlies and the polar easterlies in the upper atmosphere around latitudes 50–60 degrees. This jet stream takes a sinuous path around the globe, meander-ing first in the direction of the poles and then the Equator. As it streams pole-ward, it rises and exerts a gentle suction on the underlying atmosphere, drawing air in and up in this part of its track. The Earth's rotation adds a swirl to the converging warm temperate and cold polar air masses, and a frontal depression is born. The faster the jet stream, the more intense the depression.

Once created, the frontal depression is steered east along the track of the meandering jet stream at a speed of about 48kmh/30mph. By monitoring the position of the jet stream, using high-altitude aircraft and balloons, weather forecasters can at last predict where depressions are likely to form and the direction in which they travel.

But as one mystery is disentangled, another takes its place: the jet stream suddenly changes its wandering pattern. And until the reason for this is found, forecasts can still go wrong.

The coiled serpent

Bad weather often disrupts our well-ordered plans, but occasionally it can become so extreme and violent that it threatens our homes, possessions and even our lives. The challenge of modern meteorology is to unravel the mystery of such threatening weather conditions so that accurate warnings can be given and avoiding action taken.

One of the worst weather hazards faced by people in the tropics is the tropical cyclone. A rotating storm about 800km/500mls across, it has a distinctive cloud-free eye some 16–40km/10–25mls in diameter, around which blow violent winds. Some 80–100 of these storms form over the world's oceans each year, causing an average of $10 billion worth of damage and killing 20,000 people. They are called hurricanes in the Atlantic Ocean, typhoons in the North Pacific, Baguios in the Philippines, and simply cyclones in the Indian Ocean and around Australia.

The term cyclone comes from the Greek *kyklon*, meaning coiled serpent. The word hurricane originates either from the natives of Central America, who use the word *huracan* to describe a great wind, or it may be related to Hunraken, the Mayan god of stormy weather. Nowadays, each tropical cyclone which occurs in the Atlantic is given the name of a person, starting with the letter A at the beginning of the year and working through the alphabet.

The process by which an ordinary rainstorm is transformed into an awesome hurricane involves so many unknowns that meteorologists can only tell us what they think happens. In the North Atlantic, the starting point for a hurricane is a small, innocuous rainstorm embedded in the Trade winds off the West African coast; many of them arise in the Doldrums.

These narrow, windless zones lie as far as 3 and 7 degrees latitude on either

The driving winds of Hurricane David in August/September 1979 sent monster waves rolling with terrifying force over the low-lying Caribbean islands. Hundreds of people were killed, thousands of homes destroyed and plantations devastated. And as the hurricane moved up the coast of the United States, it spawned a string of 34 tornadoes.

side of the Equator, where the two hemispheric equatorial circulations, or "Hadley Cells", meet. For days or even weeks, sailing ships would be becalmed here as the Trade winds converged high overhead, barely ruffling the surface waters or alleviating the stifling heat. Ships would simply drift with the ocean currents, and the crew could only pray that they would eventually reach the influence of the Trades. Unfortunately for early mariners, the Doldrums shift north and south with the seasons, and their extent and precise position is quite variable.

The name of this zone derives from the Old English "dol", meaning dull. The term is, however, deceptive, for from time to time, the stillness of the Doldrums is broken by violent, squally thunderstorms, which, if they drift poleward to reach the Trades, may provide the nucleus for the birth of a hurricane. Most of these storms travel thousands of miles westward across the Atlantic and develop no further, but occasionally something triggers the storm's devastating transformation.

The cause may be a period of intense surface heating of the ocean, when surface water temperatures can exceed 27°C/80.6°F, or it may be the pumping action of high-level winds beneath which the rainstorm passes. Whatever the reason, the winds of the storm turn in on themselves and begin to form the organized circulation so characteristic of hurricanes. Spiralling bands of thunderstorms rapidly build up to the top of the weather atmosphere, and eventually the winds reach the threshold speed of 119kmh/74mph. In a typical year, more than 100 disturbances with such potential develop over the Atlantic Ocean, but only about six actually mature into fearsome hurricanes.

The movement and characteristics of tropical cyclones are closely monitored by satellites, radar and weather buoys, as well as by aircraft, whose pilots bravely penetrate the turbulent clouds defining the eye of the hurricane – the eyewall clouds – to measure wind speeds and air pressure. Even with such detailed information entered into sophisticated

computer models, predicting the course of a hurricane is immensely difficult.

Currently, in the United States, the error in predicting where the eye of the hurricane will cross the coast averages 187km/116mls for a 24-hour forecast and 399km/248mls for a 48-hour forecast. In recent years, the inability of forecasters to decrease these errors has raised questions about the inherent predictability of hurricane motion, since it appears that the eye of the hurricane moves almost haphazardly under the influence of whimsical steering currents.

It is important to improve forecasts because if inaccurate warnings are given to communities to evacuate, people are less likely to leave their homes the next time a warning is issued. For example, 50,000 residents of Miami Beach evacuated as Hurricane David approached in August 1979, but it was West Palm Beach, 80km/50mls to the east, that was struck. When Hurricane Allen threatened in August 1980, several hundred thousand people fled inland from the Texas coast as forecasters predicted winds of 322kmh/200mph; but the hurricane eventually crossed the sparsely populated lower Texas coast and the winds fell to 177kmh/110mph. Obviously a short-term warning of, say, 12 hours is more accurate, but the time needed for evacuation is often considerably longer than that.

Inside a hurricane

Ocean warming releases energy to create spiralling bands of towering thunderstorms and high winds around a cloud-free eye. In the oblique shot of Hurricane David, left, the still eye is obscured, but the immense extent and force of the swirling cloud mass is vividly brought home. One day's energy released from a hurricane and converted to electricity could supply the needs of the United States for six months.

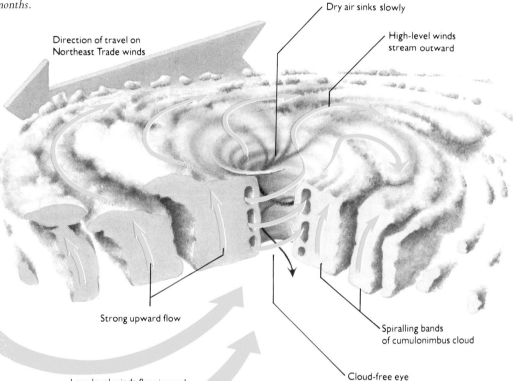

Direction of travel on
Northeast Trade winds

Dry air sinks slowly

High-level winds
stream outward

Strong upward flow

Low-level winds flow inward

Spiralling bands
of cumulonimbus cloud

Cloud-free eye

Nothing is secure when a hurricane strikes; houses are broken up, as though made of cardboard, and motorcars dumped like toys.

Perhaps coastal residents should follow the example of the Seminole Indian tribe of Florida and observe the activity of birds, rats and even alligators to know whether a hurricane is going to strike. In 1944, Florida was the target for two hurricanes. The tribe left the area as the first storm threatened, whereas the National Weather Bureau forecast it would miss Florida. For the second hurricane, the Seminoles stayed put, while the Weather Bureau issued an evacuation warning. The Seminoles were correct both times.

The energy to power and sustain a hurricane comes from the warm ocean. The Sun evaporates water and the water vapour rises and condenses to form the deep thunderclouds of the storm. As the water vapour condenses it releases a colossal amount of energy to drive the winds. Although hurricanes are incredibly powerful, some scientists in the United States have suggested that they could be tamed.

The proposition is that by using chemicals to seed the clouds in a ring some distance outside the eyewall, greater rainfall would be produced, which in turn would release more heat and generate another wall of towering clouds. It is argued that the new wall would reduce the energy feeding the original eyewall of the storm, so lessening the winds. Success was claimed during the Project Stormfury experiments with Hurricane Debbie in 1969, when winds temporarily dropped by 30 percent after seeding.

Since then, attempts have been unsuccessful, and atmospheric scientists now urge caution in case such experiments upset the weather in other areas of the world. In addition, memories of the first hurricane-taming attempt of 1947 persist. Then, at the time of seeding, the hurricane was heading away from the coast, but shortly after seeding it turned around and caused extensive damage in Georgia.

The threat to island and coastal communities from tropical cyclones comes from a combination of enormous wind speeds, battering storm surges and heavy rainfall. A measure of the potential violence of a hurricane is the air pressure in the eye. The lower the pressure, the stronger the wind speeds, the higher the storm surge, and the heavier the rainfall.

Normally, surface air pressure lies at 980–1,030 millibars (mb). In comparison, two-thirds of all hurricanes have central pressures less than 980mb, while Hurricane Gilbert, which swept through the Gulf of Mexico and devastated Jamaica in September 1988, reached a record 885mb. In consequence, it generated sustained winds of 274kmh/170mph and gusts in excess of 322kmh/200mph.

An unexpected effect of low pressure was discovered during the passage of

Hurricane Gloria in September 1985. As it swept past Long Island, New York, several hospitals reported an increase in the birth of premature babies as the pressure plunged, inducing labour.

The hurricane storm surge is a vast dome of water, often 80km/50mls wide, that comes sweeping across the coastline where the eye of the hurricane makes landfall. It is created by the low air pressure of the hurricane raising the level of the sea.

As this crest of water reaches the shallow coastal margins, waves can increase to 4.5m/15ft or more above mean water level. Improved knowledge of what storm surge height to expect from hurricanes has helped coastal communities to protect themselves by building massive seawalls and tidal barriers. After 6,000 people died in the low-lying city of Galveston, Texas, in September 1900, more than 2,000 buildings were raised 2.1m/7ft on piles to lessen their

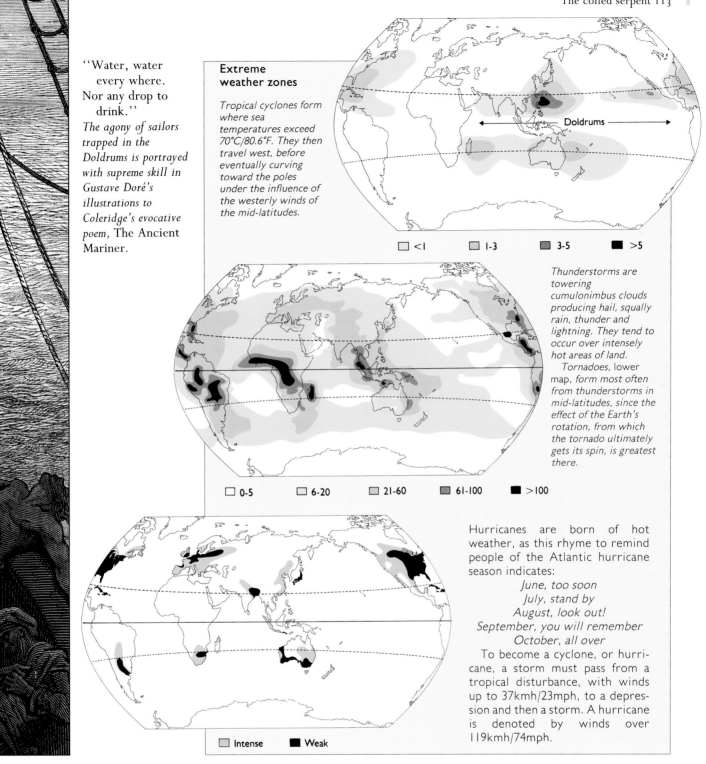

"Water, water
every where.
Nor any drop to
drink."
The agony of sailors
trapped in the
Doldrums is portrayed
with supreme skill in
Gustave Doré's
illustrations to
Coleridge's evocative
poem, The Ancient
Mariner.

**Extreme
weather zones**

*Tropical cyclones form
where sea
temperatures exceed
70°C/80.6°F. They then
travel west, before
eventually curving
toward the poles
under the influence of
the westerly winds of
the mid-latitudes.*

☐ <1   ☐ 1-3   ▨ 3-5   ■ >5

*Thunderstorms are
towering
cumulonimbus clouds
producing hail, squally
rain, thunder and
lightning. They tend to
occur over intensely
hot areas of land.
Tornadoes, lower
map, form most often
from thunderstorms in
mid-latitudes, since the
effect of the Earth's
rotation, from which
the tornado ultimately
gets its spin, is greatest
there.*

☐ 0-5   ☐ 6-20   ☐ 21-60   ▨ 61-100   ■ >100

Hurricanes are born of hot
weather, as this rhyme to remind
people of the Atlantic hurricane
season indicates:
*June, too soon
July, stand by
August, look out!
September, you will remember
October, all over*
To become a cyclone, or hurricane, a storm must pass from a
tropical disturbance, with winds
up to 37kmh/23mph, to a depression and then a storm. A hurricane
is denoted by winds over
119kmh/74mph.

☐ Intense   ■ Weak

vulnerability to enormous storm surges.

Throughout the tropics, coastal and island communities constantly fear the wrath of the tropical cyclone. A great deal of money and resources are needed to take avoiding or preventive action against hurricanes. For poor countries such as Bangladesh, faced with the threat of cyclones which funnel up the Bay of Bengal, the options are few. The situa-tion is made worse because of the low-lying nature of the delta land and its high population density. When a tropical cyclone struck in November 1970, half a million people were drowned by the 6-m/20-ft storm surge.

Once they reach land, hurricanes begin to die, cut off from their oceanic source of energy and subject to the effects of frictional drag; but they still have a sting in the tail. As a hurricane slows, the rotation of the storm becomes focused into its constituent thunder-storms and these can spawn tornadoes.

In 1967, as Hurricane Beulah struck the American coast, it created 141 torna-does in this way. Typically, 127–254mm/5–10in of heavy rain also falls at this time, benefiting farmers—but at the cost of extensive flooding.

On 22 May 1987, nearly all the 185 residents of the tiny Texas town of Saragosa were gathered in the commun-ity hall. Suddenly, a black, twisting funnel cloud dropped out of the sky with a deafening roar. Within a few seconds the community hall lay crushed; 40 people were killed and 100 injured.

The tornado that destroyed Saragosa was one of 1,000 to strike the United States each year. The death and damage tornadoes cause has generated a considerable effort to understand them. Yet, after decades of research, many questions remain unanswered and some meteorologists suggest they may even be unanswerable.

Tornadoes are seldom more than a few hundred yards across and may last only a few minutes. Nevertheless, during their existence, they can scythe a narrow path of total destruction through a community. Their small size and brief duration makes forecasting precisely when and where they are likely to strike extremely difficult.

A tornado is a funnel-shaped rotating vortex which descends from the base of a deep cloud, usually a thunderstorm, and touches the ground. Thunderstorms are fed by moist, warm, rising air and, in the most severe, the formation of a vortex provides the most efficient means for transporting this air into the base of the storm. The tornado forms when the rapidly rising air is sent spinning, usually through the influence of other air movements within the thunderstorm.

Waterspouts are essentially tornadoes over water, but the lower surface-friction of lakes, rivers and the sea means that a vortex forms more readily than it would over land. Thus many waterspouts are far less violent than tornadoes and can develop from towering cumulus clouds rather than severe thunderstorms. Nevertheless, intense waterspouts can sink ships and damage coastal communities. Waterspouts caused the collapse of the famous Tay Bridge in Scotland in 1879, leading to the Edinburgh mail train plunging into the water below and killing 75 passengers.

Destructive tornadoes occur in the mid-latitudes of both hemispheres; but fortunately, only less than one percent of thunderstorms spawn tornadoes. On average, each year 300–400 people throughout the world are killed by tornadoes, about a quarter of them in the United States, where the death toll would be much higher were it not for the tornado warnings issued to communities.

Radar has been essential in improving the ability to recognize, detect and track a thunderstorm with tornado potential. Only a few minutes warning may be given before a tornado strikes, but it is usually sufficient time for people to seek refuge. In Tornado Alley, in the US Midwest, a strip of land almost 805km/500mls long and 646km/400mls wide where tornadoes are frequent, many homes have specially constructed strong cellars in which to shelter.

A tornado poses a dangerous threat, with its powerful winds reaching as high as 402kmh/250mph, its incredible lifting power, and the sudden lowering of air pressure associated with its passage. The combined effects can make buildings explode; trucks, cars and even railway locomotives can be lifted and carried some distance; sand and gravel can pepper human bodies like gunshot, while straws can penetrate tree trunks; trees can be twisted like corkscrews and chickens can be stripped of their feathers.

However, tornadoes may have a gentler side to their nature. In May 1986, it was reported that a tornado sucked 13 schoolchildren into the air in west China and carried them for 19km/12mls before gently lowering them, unharmed, into sand dunes and scrub.

Tetsuya Fujita, Professor of Meteorology at the University of Chicago, is the undisputed "king" of tornado research. His interest was aroused in 1945 when, as a 24-year-old student, he studied the damage done by tornadoes that formed in the wake of the atom bomb explosion at Hiroshima.

In the decades since, he has followed the tracks and examined the damage done by hundreds of these twisters. Now he has devised a machine that mimics a tornado, using dry ice to provide moisture and spinning cups below a suction fan to create a swirling upward airflow.

Slowly, clue by clue, he is piecing together the causes and actions of these killers, which strike with such unexpected fury.

The United States suffers the highest number of tornadoes in the world, each bringing destruction with it. The vicious spinning winds that hit this Utah farm have sucked up debris and dust to form a dirty "sleeve" around the tornado, which, freakish and ephemeral, will vanish as rapidly as it formed.

In August 1694, a waterspout struck the little port of Topsham in southwest England. Angry locals blamed the navy for the damage it caused because they had not fired their cannon at the creature.

Bolts from the blue

Thunderstorms generate an incredible range of weather, including breathtaking lightning, ominous thunder, torrential rain, destructive hail, squally gusts of wind, and even creature falls.

A thunderstorm is formed when a strong rising current of warm, moist air changes a small cumulus cloud into a heavy, dense cumulonimbus cloud, 10–25km/6–10mls high and about 8km/5mls wide, in less than an hour. Currents of descending colder air lie adjacent to the rising air, creating a situation of extreme turbulence within the cloud. The fast-rising air supports the growth of large water droplets, ice crystals and hail, and it is collisions between these which create the electrical charges that ultimately produce lightning.

A complete understanding of how this process works still eludes us, but we know that positive electrical charges are transferred to the upper parts of the cloud and negative charges to the lower and middle parts. Eventually, an electrical difference of millions of volts is produced between these levels, as well as between the cloud and the positively charged surface of the Earth. This triggers gigantic sparks of lightning either wholly within the cloud or between the cloud and the Earth.

It took the American scientist and statesman Benjamin Franklin (1706–90) to unravel the mystery of lightning by proving it was a form of electricity. In July 1752 he flew a kite made of a silk handkerchief into a thunderstorm. Near the end of the anchoring twine, he attached a metal key, and as he moved his hand close to the key, sparks jumped the gap between it and his hand. In 1909, a Swedish scientist, Engelstad, died repeating Franklin's experiment.

With 1,800 thunderstorms happening throughout the world at any moment, lightning strikes about 6,000 times every minute. The pencil-wide lightning channel is heated to a temperature of 30,000°C/54,320°F in less than a thousandth of a second. And the rapidly expanding hot air in the channel creates a shock wave, producing the rumbling noise of thunder. Should lightning strike a moist surface, such as a tree or wall,

Some peoples still believe that thunder and lightning are weapons of the gods. This staff with the figure of Oshe, a Nigerian storm god, is shaken and prayers chanted when a destructive rainstorm threatens.

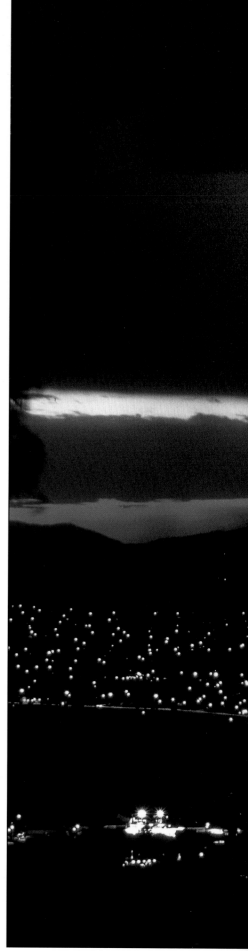

Lightning's fiery fingers, ripping through the night sky, warn of its lethal power.

the instantaneous boiling of the moisture causes such a violent expansion that it appears to explode as if a solid object, or "thunderbolt", had struck it.

Lightning seeks out the point on the ground of least electrical resistance, and trees, hills and tall buildings are most frequently struck. Indeed, the worst place to shelter from a thunderstorm is beneath a tall, isolated tree. Since it is not as good an electrical conductor as the human body, when lightning strikes a tree, it may produce a sideflash or spark to a sheltering person.

People working or playing sport outdoors also face a serious risk. Holding a metal object such as a golf club, umbrella, rifle, or garden fork increases the chances of attracting a lightning strike, which can cause dreadful burns, damage to vital organs and even stop the heart. Fortunately only some 25 percent of people struck are killed.

The safest place to be is indoors or in a vehicle. When lightning strikes a car, for instance, the current flows safely around the occupants through the metal of the bodywork, before earthing to the ground across the wet tyres. In Texas, in 1979, three passengers sitting in the open back of a truck were killed in a lightning strike, while the three inside the cab were uninjured.

Within a thunderstorm, turbulent air currents swirl water droplets and ice crystals up and down; and as the ice crystals rise and fall, they may gather alternating layers of clear and opaque ice to form hailstones. As many as 25 layers have been counted in hailstones the size of grapefruit. Once hailstones become too heavy to be supported by the storm, they fall as lethal missiles. Many lives are lost, buildings damaged and crops devastated by this "white plague".

Sometimes it is found that a very large hailstone has formed around an unusual nucleus. Two of the large hailstones that fell at Dubuque, Iowa, in 1882 contained frogs, which were still alive when the ice melted; and an exceptionally large hailstone that fell near Vicksburg, Mississippi, in 1894, contained a gopher turtle the size of a brick. At Bournemouth, England, in 1983, hundreds of hailstones 5–7.5cm/2–3in in diameter were found to contain pieces of coal. Clever detective work by meteorologists managed to pinpoint the coal merchant's yard from which powerful rising air currents had lifted the coal.

Occasionally, a truly colossal hailstone falls, such as that which landed at the feet of a meteorologist in Manchester, England, in 1973, following a single lightning strike. The hailstone, which was composed of 51 layers of alternating clear ice and bubbles, fractured, but the pieces weighed 1.5–2kg/2–4lb. There were no impurities to suggest that the ice block had been formed from waste water ejected by an aircraft, so, in some unknown way, it seems that lightning had formed this huge ice meteorite.

The weather can surprise us all, but sometimes it can truly amaze us, dropping strange objects and creatures from the sky, often a great distance from any place in which they could have originated. Increasingly such incidents are attributed to the thunderstorm and its offspring the tornado or waterspout, whose power and freakish behaviour are still largely unexplained.

Creature storms

Plagues of frogs, fish, crabs, eels, tadpoles, sea shells, hazelnuts, snails, worms and even maggots have fallen from the sky. To the bewildered people on whose heads they have descended, each incident is an absorbing mystery.

Scientists attempting to explain such occurrences are increasingly attributing them to thunderstorms. The 96kmh/60mph rising air currents of a thunderstorm, or the powerful suction of the tornado or waterspout it may spawn on rare occasions, provide the lifting force needed. As the storm passes over a shallow pond or stream, the entire contents may be sucked into the cloud and carried a long way before being dropped.

But some events remain difficult to explain and are so bizarre that they have passed into legend. In the fourth century AD, chroniclers reported a three-day fall of fish in the Chersonesos district of Greece. There were so many fish that the roads were blocked and people were unable to open their doors. Not surprisingly, the town stank for weeks after.

A rain of fishes in Saxony in 989, and a fall of frogs or toads in 1355, are among the incidents illustrated in the Chronicle of Marvels and Spectacles *published in 1557. These mystifying events were attributed to supernatural causes.*

Anvil-shaped cloud points
in the direction the storm
is moving

Strong up currents

Repeated circulation
of ice crystals forms
hailstones

Strong down currents

Heavy rain or hail

Warm air drawn up into
the base of the storm
may also catch up
small creatures

Lightning discharges flash
between the positively
charged Earth and
negatively charged cloud

*Benjamin Franklin's famous
experiment with a kite proved
that thunderstorms generate
electricity. Louis XVI of
France, trying the same thing,
passed a current through 200
monks who were holding
hands.*

*The commonest form of storm
cloud is the black, anvil-
headed cumulonimbus, which
may reach 16km/10mls into
the sky and stretch for 64–
80km/40–50mls. Within its
towering height, violent air
currents rise and descend
causing torrential rain and
hail to fall; and lightning is
discharged between the Earth
and the cloud.*

Weather lore and legend

Forecasting the weather is fraught with difficulty because so much about how the atmosphere works remains a mystery. Even though current forecasts are based on a global network of highly sophisticated monitoring instruments and employ incredibly complex computer-based models of the atmosphere, their accuracy is often criticized as less than perfect. And this dissatisfaction, especially with local forecasts, has tempted many to revert to using weather lore rather than relying on high-tech predictions.

Weather lore consists of signs, sayings and rhymes indicating how the weather may change. These are part of the folk tradition of every culture and have been passed down the centuries by word of mouth. Some are hardly more than whimsical rhymes, with little chance of being reliable:

If November ice will bear a duck,
Most of winter will be slush and muck.

When spiders weave their webs by noon,
Fine weather is coming soon.

Others are based on careful observation of the way plants and animals behave in relation to weather changes.

The aspect of the sky, including the popular saying "Red sky at night, shepherd's delight . . ." is often regarded as an indication of the weather to be expected. And a similar rhyme:

A rainbow in the morning is the shepherd's
warning;
A rainbow at night is the shepherd's
delight.

is firmly based on accurate observation, for a rainbow occurs when raindrops lie opposite the Sun: in the evening the rainbow is seen in the east, in the morning in the west. With heavy rains usually being brought by westerly winds in the Northern Hemisphere, a morning rainbow indicates that bad weather is on the way, whereas an evening rainbow is a sign that the rain is passing away.

During a spell of wet weather some claim: "When you can see enough blue to make a Dutchman's coat (or a sailor's trousers), the weather will clear." Again, observations do indeed suggest that, after a prolonged period of rain, blue sky seen through breaks in the

Elusive and beautiful, the rainbow seems always to move away as one approaches. Small wonder that the ancient Greeks saw it as Iris, messenger of the gods, and the Vikings as the bridge leading to the heavenly city of Asgard.

The belief that a red sky at night foretells a fine day is ancient—and correct. We now know that the red glow is caused by dust in the air scattering the Sun's rays, a sure sign of dry weather. Early peoples were able to gain an understanding of many weather patterns simply by observation.

cloud is a reliable indication that the cloud has thinned, and better weather should follow.

Some animals are considered especially sensitive to weather changes. High-flying swallows are said to presage good weather, and, in some parts of Europe, frogs were once used in homes as cheap substitutes for barometers. The frog was kept in a glass jar half-filled with water, in which was placed a ladder. When rain was on the way, the frog was believed to stay in the water, croaking noisily. If it sensed clearing weather, it was said to ascend the ladder; the higher it climbed, the better the weather to come.

Many plants have been observed to close their petals when moisture in the air increases and rain is on the way. As a result, some plants, among them the Scarlet Pimpernel, *Anagalis arvensis*, are known colloquially as "the poor man's weather glass".

A surprising number of people seem to be sensitive to an approaching storm and claim that they can foretell weather changes because they can "feel it in their bones". As a storm gets nearer, it causes air pressure to fall and the air to become moist, which makes bones and joints ache, scalps itch and noses irritate. The recognition that hair stretches and becomes limp when the air is damp meant that the American Zuni Indians, in days gone by, had a convenient weather sensor: "When locks turn damp in the scalping tent, it will rain on the morrow."

Weather lore is most successful in forecasting changes a few hours ahead. As the time-scale lengthens, so does the tendency for a prediction to fail. Too often weather lore is based simply on wishful thinking. For instance, severe winters are mistakenly foretold in the belief that nature prepares itself by supplying extra food for birds and animals in the form of abundant berries, nuts and acorns.

Weather lore, like modern-day forecasting pays particular attention to predicting weather extremes. Imagine how society would benefit if, instead of having to attempt to warn of such

extremes, they could be eliminated. If the fury of storms could be tamed and droughts ended by abundant rainfall, then lives would be saved and devastation prevented. Early unsuccessful weather control attempts were limited to magical rites, prayers, sacrifices, the ringing of church bells and firing of cannons and small rockets.

A major breakthrough came in 1946 when American scientists discovered that seeding a cloud from an aircraft with chemicals, such as silver iodide or dry ice, could produce rainfall. In the

excitement following the discovery, many people believed that disastrous droughts could be ended and that the same technique could be used to reduce the fury of hurricanes, hailstorms and electric storms. Irving Langmuir (1881–1957), the American Nobel Prize winner, optimistically suggested there was "a reasonable probability that in one or two years man will be able to abolish most damage from hurricanes."

Scientists had underestimated the complexity of the atmosphere, however. Many experiments failed to

A novel Storm Warning System was shown at the Great Exhibition in London in 1851 by Dr Merryweather of Whitby in Yorkshire. When storms threatened, leeches in the bottles became active and a bell rang. The doctor urged the government to set up a network of his devices along the coast, but the scheme was rejected.

Measuring the atmosphere

Cameras and sensors on geostationary satellites, some 36,000km/22,500mls above the Equator, send back images of the atmosphere. Meteorologists use the data to build up a picture of how the world's weather might develop.

It is accurate measurement that distinguishes even the most successful interpretation of weather lore from the science of meteorology. The first instruments to measure the atmosphere were thermometers and barometers containing mercury and the anemometer to measure wind speed. Later, instruments were sent up in balloons to obtain data from high in the atmosphere. Now the Earth is scanned by satellites, and a stream of information sent to weather stations, using cameras, radio sonde and electronics.

produce the expected outcome, and some even produced unwelcome results, highlighting our limited understanding of what drives the weather machine. Further, new research found that a change in the weather in one part of the world might eventually affect the weather in a far distant region, so making apparent the dangers inherent in interfering with the weather. Scientists now believe we must strive toward a better understanding of the puzzling processes that make up the weather before we can attempt to alter it.

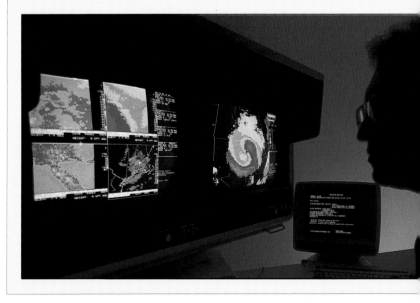

El Niño:
currents in combat

Each year, during December, a gentle current of warm water spreads through the South Pacific Ocean. It moves toward the coast of Ecuador and northern Peru above the cold, deep, north-flowing Peru (Humboldt) Current. Long ago, fishermen called this warm-water invasion "El Niño", meaning "the boy" in Spanish. It also translates as "the Child", a name deriving from El Niño's habit of arriving round about Christmastime.

Every three to four years, for no known reason, a far more pronounced and extensive ocean warming occurs throughout the equatorial central and eastern Pacific Ocean. This strong and long-lasting warming (typically over a period of 14–18 months) has such profound consequences for the world's weather that today the name "El Niño" is reserved for these major events.

Normally, in December, air pressure over the southeast Pacific is high, with associated sinking air, and low over Indonesia, indicating rising air. When an El Niño event occurs, this situation is reversed. Air pressure over the southeast Pacific drops as the region experiences rising air, while over Indonesia and Australia the pressure increases and air currents sink. Clearly, El Niño is part of a giant ocean-atmosphere seesaw circulation, whose pressure pattern was first recognized in the 1920s by Sir Gilbert Walker, known as the Walker Circulation or Southern Oscillation.

In the Pacific Ocean, the northeast and southeast Trade winds usually converge along the equatorial zone known as the Doldrums. What winds do occur here, blow westward, dragging with them the surface waters, which are significantly warmer than the deep waters beneath. This causes a rise in sea level in the western Pacific of 30–70cm/ 1–2ft, and a similar lowering in the eastern Pacific. This allows cold water

The devastating influence of El Niño, whose onset we can neither understand nor predict, is felt worldwide. As these bleached bones show, even the camel, with its celebrated endurance in waterless conditions, cannot withstand the harsh drought and famine that recent El Niño events have brought to countries in eastern Africa.

to well up from the depths along the coast of South America.

Rich in nutrients and plankton, these cold waters of the Peru Current make an excellent fishing ground. Indeed, the enormous shoals of small anchovies, used commercially to produce fish meal for animal feed, form the basis of the fishing industry of Peru and Ecuador, which in 1971 was the world's largest.

When El Niño occurs, there is a surge of warm surface waters eastward across the Pacific; associated with this is a change in the direction of the equatorial easterlies, which reverse to become westerlies. The winds help the flow of surface waters to the east, which leads to piling up of warm waters along the South American coast, and the suppression of the upwelling of deep water. The lack of this nutrient-rich cold water spells disaster for the fishing industry. During the 1972 El Niño, it was almost ruined; in 1982–83 the usual catch was halved.

Such changes in the patterns of air pressure and wind have major implications for the weather throughout the

Pacific region. Usually, high pressure over the central Pacific and the South American coast produces extremely dry weather, while low pressure brings cloud and heavy rainfall to the western Pacific. When El Niño occurs, this pattern is reversed. Even the area experiencing tropical cyclones is altered. In 1982–83, 25,000 people were left homeless when Tahiti and nearby islands, normally untouched, were hit by six devastating cyclones.

During the 1982–83 El Niño events, record droughts struck parts of Australia—leading to bush fires and terrible dust storms—and Indonesia, where famine resulted. At the same time, heavy rainfall in the eastern and central Pacific regions produced catastrophic flooding and mudslides in the Andean highlands of Ecuador, Peru and Bolivia.

While El Niño persists, huge amounts of extra heat and moisture, added to the atmosphere by evaporation of the warmer ocean, affect atmospheric circulation worldwide. In 1983, winter storms, which usually cross the northern

coast of North America, were forced south, bringing rain and stormy weather to California; Florida and Cuba also had abnormally wild, wet weather. And the drought in the American Midwest in 1988 was blamed on the aftermath of the 1986–88 El Niño, which diverted rain-bearing clouds away from the area.

Recent studies highlight the fact that, because of related ocean warmings and atmospheric seesaws in equatorial regions of the Atlantic and Indian Oceans, El Niño's influence extends to other parts of the world. The severe drought in southern Africa in 1982 was the result of an El Niño event, as was the worsening of the droughts in the Sahel and Ethiopia during the 1970s and '80s.

Although we now comprehend the mechanisms of such occurrences, exactly what triggers El Niño remains undiscovered. If we could increase our understanding, we might be able dramatically to improve weather forecasts for months or even seasons ahead and so be able to counter the most destructive consequences of its onset.

In a normal year, a silvery harvest of thousands of tons of anchovies is netted by fishermen off Ecuador and Peru. When El Niño occurs, the cold Peru Current does not reach these shores and the fishing fails. Not only do people suffer, but birds and animals that depend on the fish starve.

Normal
Cold, nutrient-rich currents well up on the west coast of South America. Equatorial surface winds, blowing westward, pile up warm surface waters in the southwest Pacific. Rising air over Indonesia brings rain.

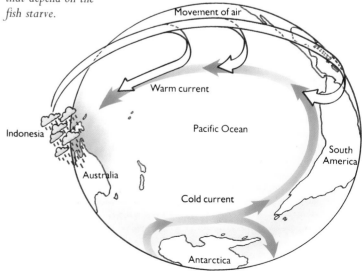

El Niño

Air movements are reversed. Warm water flows eastward from Indonesia down the coast of South America, suppressing the cold current. Rising air over the Americas results in storms and flooding.

El Niño upsets the usual pattern: drought rules where there should be rains, and floods where there should be sunshine.

In Australia, vital top soil disappears in dust storms, while in California people's homes are awash.

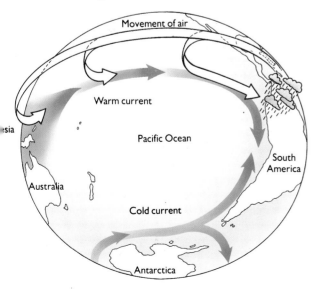

Movement of air

Warm current

Pacific Ocean

South America

...sia

Australia

Cold current

Antarctica

Shifting desert sands

In around 150BC, the Romans colonized North Africa and it became the granary of the Empire, producing some 500,000 tons of wheat a year during Julius Caesar's time. More than 600 cities flourished there, but over the centuries they have been overwhelmed by the northward shift of the Saharan sands, their ruins bleak reminders of man's puny struggle against the overwhelming and, as yet, incompletely explained, spread of the desert.

"Desertification", the process whereby vast areas of productive land, world-wide, are being turned into useless desert, with little or no water, soil or vegetation, has proceeded at an alarming rate in recent times. Between 1882 and 1952, the percentage of the Earth's land surface classified as desert rose from 9.4 to 23.3. In 1984, the United Nations Environment Program stated: "35 percent of the world's land surface is at risk . . . each year 21 million hectares is reduced to near or complete use-lessness." Why are such large areas in danger of such degradation? Is it due to a change in global climate because of the warming of the Earth's atmosphere? Or is it a consequence of human mismanagement of the environment?

The key factor determining a desert is rainfall. An area receiving less than 100mm/4in of rain a year is classified as true desert. Less than 250mm/10in denotes an arid area, while less than 500mm/20in is semi-arid and usually supports some form of grassland. Rainfall is a product of global air currents or winds, which, in turn, are influenced by ocean currents; and all are ultimately determined by the Sun's heating of the Earth's surface.

At the Equator, where the Sun's power is greatest, warm moist air rises into the upper atmosphere, where it cools, with cloud formation and subsequent rainfall. The air currents continue

Deluged in sand, an abandoned house on the edge of the Salah Oasis in Algeria epitomises the human struggle against the relentless encroachment of the world's deserts. Year by year, once fertile lands are engulfed by the dunes' progress. And as the mysterious forces of nature work to deflect life-giving rains from arid regions, hungry generations of people and their beasts destroy the little vegetation that remains.

The basic need of desert-living plants and animals is to conserve water and so to avoid overheating, and they have developed many ingenious adaptations to enable them to achieve this. Some animals are active in the cool of the night; others seek shade underground. The Kalahari ground squirrel puts up its tail to act as a parasol. The American jackrabbit, Saharan fennec fox and Gobi hedgehog radiate heat from their large ears. And in the Namib, darkling beetles congregate at night to exploit the sea fog that condenses on their legs and runs down into their mouths.

Plants, too, have contrived intricate means of getting and retaining water. The creosote bush of the American southwest has a root system that penetrates deep and wide to suck up all available water; while the kokerboom, *Aloe dichotoma, left,* of southern Africa stores water in its thick trunk and permanent succulent leaves.

to flow north and south to cooler latitudes and eventually descend, some 1,500km/930mls from the Equator, in the region of the Tropic of Cancer in the Northern Hemisphere and Capricorn in the Southern. The air has by this time shed most of its moisture, so no rain falls on the warm lands below. In addition, as the air passes over the land, any moisture present is sucked up and returned in the air flowing back to the Equator. This is why deserts are located in and around the tropics.

When rainfall in an already dry area decreases over a number of years, drought ensues. During the last ice age, 20,000–10,000 years ago, a great ice-sheet oscillated back and forth over northern Europe, bringing alternate periods of regular rainfall and drought to the Sahara. During the present inter-glacial, the ice has retreated and the rainbelts, too, have moved north, leaving the Sahara ever drier.

In more recent times, the American Midwest suffered severe droughts in the 1890s and again in the 1930s, when a vast area of the prairies was reduced to a dust bowl. Economic pressures led to over-grazing, exacerbating the natural catastrophe and resulting in permanent desert areas.

In Africa, too, drought has devastated the semi-arid scrub and savanna bordering the sub-Sahara, and the desert has spread southward. The Sahel belt now runs right across the continent, and drought has hit southern Sudan and Ethiopia also, bringing the horrors of famine to millions of people.

The drought in the Sahel seems to run in roughly six-year periods: 1910–15, 1941–49 and 1968–74. The most recent drought began in 1982 and appears to be ongoing. Records show a decrease in rainfall of 15 percent since 1920; in the period 1960–70, the drop was a dramatic 65mm/2½in, an annual shift southward of the desert climate of 9km/5½mls. At the same time, Mediterranean lands have been receiving increased rainfall during the winter.

Some climatologists link the droughts in Africa with the varying strength of the fast-moving, high-altitude Polar Front Jet Stream. When the jet stream is weak, it suppresses the northern extent of the rains in Africa and India, and drought results. Other experts implicate El Niño, the periodic, abnormal warming of the oceans' surface water.

Scientific evidence shows that today's deserts were once much wetter. For instance, deep canyons running through the Negev Desert in Israel indicate the former presence of great rivers, and pollen analysis of sediments reveals that rich vegetation once grew there. And rock paintings, dating back 2,000–5,000 years, in the ancient caves of the Tassili plateau in the central Sahara, depict a pastoral people, with herds of piebald cattle, and wild animals such as gazelle, hippopotamus and giraffe. Evidently this was once well-watered savanna.

The Roman experiment in large-scale agriculture in North Africa lasted only a few hundred years. Forests were cut down for timber and to make way for agriculture; overcropping exhausted the soil, and overgrazing denuded the vegetation cover, so the topsoil blew away. The once fertile land became desert. But nature, too, was playing a part: by AD 250 Saint Cyprian, Bishop of Carthage, had recorded that the rainfall had diminished and once-abundant springs barely gave a trickle.

The ecological mistakes of yesterday are still being repeated, and added to them is the daunting problem of over-population. But, at the heart of the matter are the vagaries of climate. Its unpredictable fluctuations coupled with human overexploitation of the world's resources form a deadly combination.

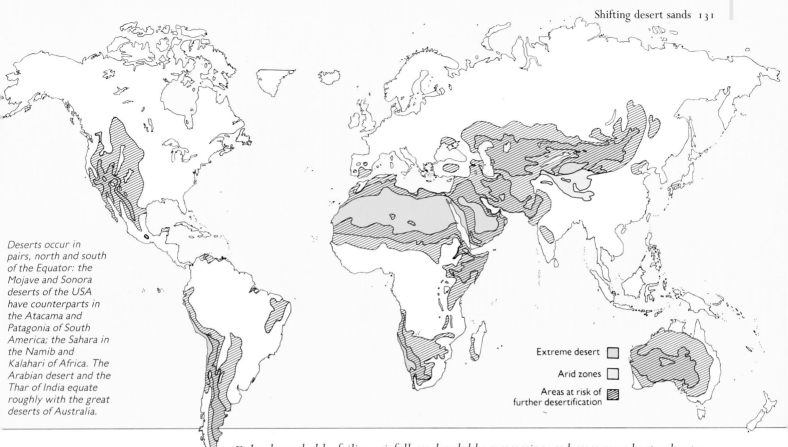

Deserts occur in pairs, north and south of the Equator: the Mojave and Sonora deserts of the USA have counterparts in the Atacama and Patagonia of South America; the Sahara in the Namib and Kalahari of Africa. The Arabian desert and the Thar of India equate roughly with the great deserts of Australia.

Extreme desert ☐
Arid zones ☐
Areas at risk of ▨
further desertification

∞ *Lands parched by failing rainfall are denuded by overgrazing, and areas around natural water sources are often the first to be turned into desert.*

The climate conundrum

In the 3,500 million years since life began, the Earth's climate has fluctuated substantially. Nevertheless, it seems to have remained comfortable enough to ensure the survival of plants and animals, despite the fact that in this immense span of time the Sun has radiated ever more heat to our planet.

Scientists speculate whether some mysterious self-regulatory mechanism or "Gaia Effect" has been at work, influencing the planet's climate, ecology and evolution. Yet even before we have a chance of solving this mystery, the human inhabitants of the Earth have embarked on their own experiment in climate change, whose outcome has enormous destructive potential.

At the present time, the Earth is warming up. A hotter planet will be more susceptible to disruptions in agriculture and fisheries, and melting of the polar ice caps will lead to the flooding of many coastlines and cities. Eventually, at least in theory, the planet might become uninhabitable.

Charting the long-term future of global climate may be as speculative as crystal-gazing, but what is certain is that our climate reflects the balance between energy received from the Sun, the amount the Earth uses and stores to maintain a life-sustaining environment, and the amount of energy that is returned to outer space.

The Sun's energy, in the form of short-wavelength light, is absorbed by atmospheric gases, clouds, oceans and land. It is thus converted into heat, which warms the planet, drives the winds, evaporates water and ultimately produces rainfall.

The amount of heat temporarily stored within the Earth's atmosphere is critical in establishing global temperature. Gases present in the atmosphere in relatively small amounts, which include carbon dioxide, chlorofluorocarbons

Virgin rain forest, a world of mystery, excitement and unimagined diversity, is the very essence of the magnificence of nature. Curiously it is also at the heart of the healthy life of our planet, for without the balance the rain forests so tenuously maintain between carbon dioxide and oxygen in the atmosphere, our Earth is destined to wither and die.

(CFCs), methane, ozone and nitrous oxide, play a key role in determining the amount of energy stored. Together with water vapour, they allow the Sun's energy to pass through the atmosphere unhindered but absorb a significant portion of the long-wavelength energy that our planet gives off to outer space.

Like the panes of glass in a greenhouse, the gases let light in, but trap the heat; through this process, known as the "Greenhouse Effect", they are vital to maintaining our comfortable global temperature. Without the existing levels of greenhouse gases our planet would have a chilly global surface air temperature of around $-17°C/+1.4°F$. And if the Earth had the amount of greenhouse gases found on Venus, we would suffer global temperatures in excess of $500°C/932°F$. The current increase in levels of greenhouse gases is too small to transform Earth's climate into a Venusian-hot climate, but it will nevertheless cause a significant rise in temperatures.

Carbon dioxide is the most important greenhouse gas: it is responsible for about half of the expected warming over the next century. In the past 100 years, as the levels of all greenhouse gases have increased, global temperature has risen by about $0.5°C/0.9°F$. Natural fluctuations in the climate have masked some of this increase, but by the late 1980s the warming was confirmed as the decade experienced the six warmest years that have occurred this century.

Most scientists agree that if greenhouse gases continue to increase, by the middle of next century, global warming is likely to be in the range $1.5–4.5°C/2.7–8.1°F$. This may seem a small amount, but the planet will not have been so warm since the period before the last ice age.

Global warming is expected to change regional climate and rainfall patterns, so displacing the best areas of agricultural production. In particular, soils in the main wheat- and corn-growing areas of the mid-latitudes, especially the American Midwest, may become drier with a consequent decline in crop yields.

Since the world food market depends

largely upon surpluses from this area, the implications may be devastating for low-income, food-importing developing countries. The survival of millions of people in the Third World may depend on the ability of farmers and plant breeders in the developed countries to adjust to the changing climate and to maintain production in a less equable climate.

Decreased crop yields due to climate changes may be partly offset by the direct effects of higher concentrations of carbon dioxide in the atmosphere. Laboratory experiments indicate that a doubling of carbon dioxide concentrations enhances photosynthesis (the process by which plants use light-energy to convert water and carbon dioxide into food). It also increases the growth and yield of crops such as wheat,

The burning of vast areas of dense tropical forest, in South and Central America and Southeast Asia, not only releases carbon dioxide but reduces the amount of vegetation available to remove large quantities of the gas from the atmosphere.

Belching chimneys and coolings towers of industries using fossil fuels—coal and oil—currently produce an average of more than a ton of carbon for every person on Earth each year. The emissions also contain nitrous oxide and cause photochemical smogs of oxides of nitrogen and hydrocarbon; sunlight acting upon these creates eye-irritating ozone.

The "Greenhouse Effect"

The gradual warming of the Earth's atmosphere is the result of several processes, which are shown in the diagram. The movement of carbon dioxide, the most important of the greenhouse gases is shown, as well as the carbon cycle and the passage of carbon through the world ecosystem.

Of the other significant greenhouse gases, the most notorious are currently chlorofluorocarbons, CFCs. First developed in the 1930s, their most common application is as propellants in aerosol sprays, in refrigerator cooling systems and in blowing plastic foams.

Nitrous oxide, or laughing gas, comes from fossil fuel burning, and from nitrogen-based fertilizers and animal wastes. Levels are increasing at a rate of around 0.2 percent a year. The guts and droppings of ruminant animals, such as cattle and water buffalo, also produce methane, as does bacterial activity in swamps and rice paddies and waste landfill sites. Levels are rising at about 1 percent a year.

Light energy from the Sun

Earth's atmosphere

Methane

CFCs

Nitrous oxide

Some light energy is reflected back into space

Greenhouse gases increase the atmosphere's ability to trap heat

Most light energy passes through the atmosphere

The Earth absorbs light energy and radiates heat energy

Some heat energy is kept in by the atmosphere

Most light energy escapes

Fixed carbon is trapped as coal and oil

Photosynthesis by plants and phytoplankton fixes atmospheric carbon

Burning of fossil fuels returns carbon to the atmosphere rapidly

Natural processes return carbon to the atmosphere relatively slowly

The amount of CO_2 in the air, measured at Mauna Loa, Hawaii, in 1960 was 270ppm, much the same as in the mid-1800s. By 1980 it had increased to some 340ppm. Estimates based on these data predict a doubling of CO_2 in the air by the year 2080.

Carbon dioxide: parts per million

340

335

330

325

320

315

310

1962 1964 1966 1968 1970 1972 1974 1976 1978 1980

soyabean and rice by around one-third.

Further, a higher carbon dioxide concentration reduces the size of the stomata (the tiny leaf pores through which gases and moisture pass), leading to an improvement in the efficiency with which plants use water. It seems, then, that plants could tolerate considerably drier conditions and still produce very similar yields.

Much of our knowledge about the direct effects of carbon dioxide increases comes from the safety of the scientist's laboratory, not the farmer's field. How crops will fare in a carbon dioxide-enriched world, with strong competition from weeds, pests, diseases and changeable weather conditions, is much less certain.

Global warming will make the water in the upper layers of the oceans heat up and expand, causing a rise in sea levels worldwide. Melting mountain glaciers will add to this rise. Global sea levels have already risen by about 15cm/6in this century in response to the small greenhouse warming that has already taken place.

Predictions of what increase in sea level to expect vary considerably. It could be a catastrophic 4.5–5.5m/15–18ft if the West Antarctic ice sheet were to disintegrate and melt within a few centuries. However, conservative estimates suggest a possible 90cm/3ft rise toward the end of the twenty-first century; but even this would have extremely grave consequences for low-lying islands and coasts.

The first warning about the Greenhouse Effect came in 1896 from the Swedish scientist Svante Arrhenius (1859–1927) but it was ignored. Now, although global warming can be reduced or delayed, it cannot be stopped because

The enigma of global warming

One of the most baffling aspects of the Greenhouse Effect is exactly how it will affect the Earth. Scientists cannot agree, and speculation is rife and sometimes alarmist. At the present rate of increase in global temperatures, the Earth could be at least 2°C/3.6°F warmer by the middle of the twenty-first century. Melting of the polar ice caps could produce a 1-m/3-ft rise in sea levels, resulting in the flooding of part of great and populous cities such as New York, London and Tokyo. Industrial complexes and prime agricultural land along low-lying coasts could be inundated, the Maldives, a collection of 1,196 atolls in the Indian Ocean inhabited by some 177,000 people, would virtually disappear.

There would also be far-reaching changes in climate, not all of them bad in themselves. Some areas of the world, notably in higher latitudes, would warm up significantly and rainfall would increase, permitting crops to be grown in, for instance, northern Canada and Scandinavia, where previously it has been too cold for agriculture. But other regions that are now productive, such as the grain belt of the American Midwest, would experience a marked drop in rainfall, and most continental interiors would suffer from drought.

Peasants in low-lying countries such as Bangladesh already know the despair of having their farms and homes flooded when monsoon storms drive the sea far inland. They are among the first people who will suffer the total loss of their land if the melting of the polar ice caps, due to global warming, raises sea levels significantly and permanently.

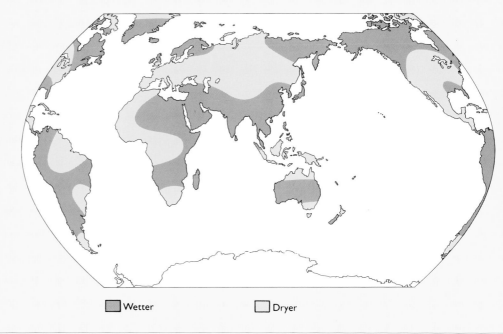

Wetter   Dryer

---

the present concentrations of greenhouse gases have already committed the planet to a further rise of 0.5–1.0°C/0.9–1.8°F.

Worldwide, it is the burning of coal that contributes most to the Greenhouse Effect through the release of carbon dioxide, and it will continue to do so. There are no magical technical solutions available, but reducing energy consumption through better use of energy in factories, homes, offices and on the roads, and replacing coal with non-fossil fuels (hydroelectricity, solar, wind and tidal power, or nuclear power) could delay the most severe effects by several decades.

The protection of tropical forests and widespread planting of broadleaved trees would also be a major step, since their leaves absorb carbon dioxide from the atmosphere. The elimination of CFCs and control of other greenhouse gases could also be achieved.

The time gained by these measures would provide a breathing space before the worst warming occurs. Farmers could begin to expand irrigation, to develop new crops and seed varieties and even to plan the movement of farming operations away from the worst affected areas. Similarly, coastal communities would have time to build or strengthen barriers against rising water levels and to modify port facilities.

Since the early 1970s, scientists have been expressing serious concern that pollution may not only cause global warming, but may have an even more sinister effect—the depletion of the ozone layer. Ozone is a form of oxygen which has three atoms in each molecule (O_3), in contrast to the usual form of oxygen which has two atoms (O_2). It is created in the upper atmosphere by light from the Sun splitting ordinary oxygen molecules, which then recombine to form ozone.

Most of the world's ozone lies in a layer 13–24km/8–15mls above our heads. Although ozone constitutes less than one-millionth part of the gases in the atmosphere, and spread evenly around the world at sea level would be only 3mm/$\frac{1}{8}$in thick, it is essential for the wellbeing of life on our planet. It acts as a protective shield for the Earth, absorbing large amounts of the Sun's very short-wavelength, or ultraviolet light, which would otherwise harm animals and damage plants.

The anxiety of scientists went largely unnoticed until May 1985, when the British Antarctic Survey Team at Halley Bay, on the Antarctic coast, reported a massive thinning of the ozone layer overhead. They blamed pollution, principally by chlorofluorocarbons (CFCs), which were already causing concern because they contribute about one-fifth to the global Greenhouse Effect. The thinning of the ozone layer or ozone "hole" would doubtless have been discovered earlier had not the computer analyzing the US Nimbus satellite measurements been programmed to exclude very low readings as errors.

The build-up of CFCs has little effect on the ozone layer until some—unknown—critical point is reached. The Antarctic appears to provide a situation where that threshold is reached more quickly than in other parts of the world, consequently a small increase in the amount of CFCs in this region causes a sudden, large depletion of ozone.

The ozone hole over the Antarctic lasts for three months each year, beginning in September with the return of the Sun after the long polar night: springtime in the Southern Hemisphere. During the winter, a ring of exceedingly strong winds encircles the Antarctic, separating the air over that frozen continent from the air over the rest of the world. In this isolated zone, at temperatures around $-85°C/-121°F$, complex chemical reactions involving CFCs and nitrous oxide take place. They lead to chlorine compounds contained

The ozone hole is largest in spring because strong winter winds isolate Antarctica, causing CFCs to build up in the atmposhere. The CFCs react with oxygen to form chlorine compounds, which destroy ozone.

Garbed like a surgeon to avoid contamination, a glaciologist saws off a piece of an ice core drilled from an Antarctic snowpit. Analysis of the ice enables scientists to check for chemical and radio-chemical trace elements. The data provides evidence of past climatic changes and of current levels of atmospheric pollution.

by the CFCs destroying the ozone molecules; one chlorine molecule can remove 100,000 ozone molecules in a year or two.

A depleted ozone layer results in higher levels of damaging ultraviolet light reaching the ground. Normally, ultraviolet light causes sunburn, but at greater levels it can lead to skin cancers: a 5 percent decrease in global ozone could generate three million new cases of skin cancer each year. Studies in Australia and New Zealand, countries close to Antarctica, indicate that an equivalent increase is already taking place. Ultraviolet light also interferes with the body's immune system, which fights off infections, and it causes cataracts, one of the main forms of human blindness, as well as several serious eye complaints in cattle.

Markedly increased ultraviolet levels would threaten the phytoplankton, the single-cell plants upon which almost all marine life ultimately depends for its sustenance. Certain crops, particularly the important food crop, soya bean, could suffer greatly reduced yields, and the larvae of many fish, which live near the sea surface may be killed. Also, with less ozone absorbing ultraviolet light,

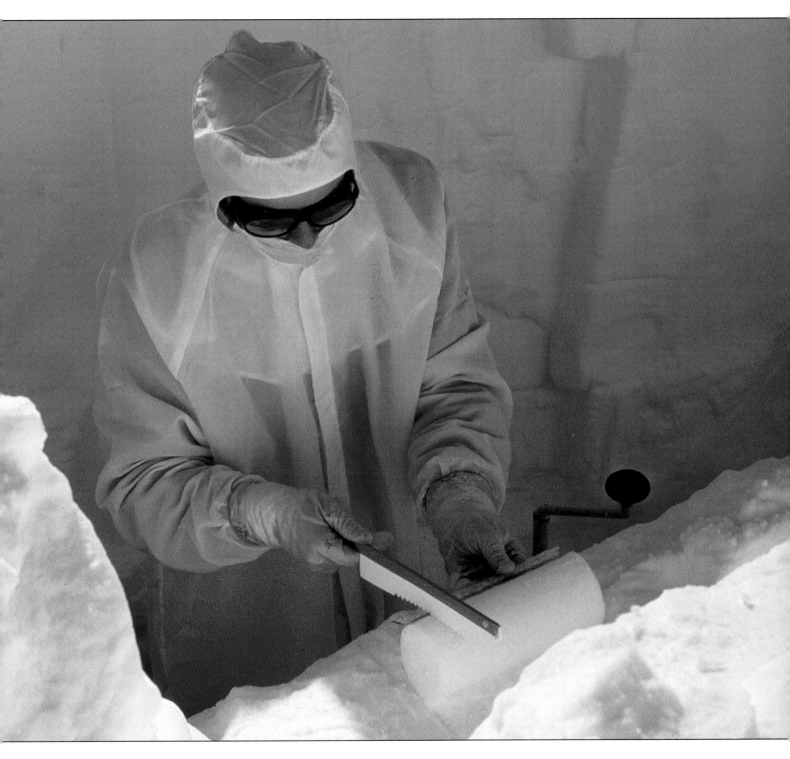

the upper atmosphere will cool and this may cause changes in weather patterns and climate.

The first action against CFCs was taken in 1978, when the United States banned their use in non-essential aerosol sprays. Other nations have been slow to follow, but since October 1987, when parts of the lower Antarctic stratosphere were being depleted of virtually all ozone, a large number of countries have finally committed themselves to cut back on CFCs. A United Nations agreement to phase out CFCs totally by the end of the century now awaits ratification.

But CFCs can remain in the atmosphere, releasing chlorine and destroying ozone for more than 100 years. Only time will tell whether the international action has been taken quickly enough to prevent a global reduction in the ozone layer. The threshold for a substantial depletion of ozone does not yet seem to have been reached for the world as a whole, but in 1989 there were signs that the critical point had already been reached in the Arctic. And the atmosphere may have yet other surprises in store for the Earth's inhabitants.

The face of the Earth

There is a perplexing blind spot within the global view of the inhabitants of the industrialized world. This "visual defect" is a bizarre form of separation between our potential and actual awareness of the wonders and mystery of the world around us—a chasm between possibilities and reality.

In the last decade of the twentieth century, we are living in the age of the "global village". Rapid, easy communication by facsimile machines, satellites, and planet-wide television linkages means that we can experience events on the far side of the world as they occur, in full colour and stereo sound. The June 1989 massacre in Tiannenmen Square, Beijing, could, so to speak, shed blood on our sitting-room carpets. We have set up international agencies that oversee human activities—security, health, commerce, aid—on a whole-world basis. Our economies are intrinsically internationalized: if

Wall Street sneezes, the City of London goes down with double pneumonia; a war in the Middle East cuts down oil production and the effects spread worldwide to hit every industrialized society.

Such is the potential of our world view that we should be in a position to enjoy a perspective of unparalleled breadth and detail concerning the planet on which we live. We have the means to scrutinize every spot on the world's surface. We should all be planet-scale natural geographers.

In fact, nothing could be farther from the truth. Somehow, ease of access to moving pictures from around the Earth seems to have dulled the immediacy of our world vision. Repetitive, barrage coverage of international news has blurred the precision with which we are able to distinguish place from place, discriminate landscape from landscape. Surveys of supposedly well-informed citizens in both

Planet Earth from space is mysterious, majestic and wonderful.

the United States and the United Kingdom have shown a frightening lack of knowledge concerning the geographical context and location of key areas of the world. At the time in 1988 when every TV news report had details of the ''tanker war'' in the Persian Gulf, few US high school students, when questioned, could even place the strategically vital stretch of water in the right continent, let alone locate it on a map.

Perhaps the very ease of access that we enjoy to carefully packaged visual images of the world is partly to blame for this geographical blind spot in our interest and understanding. Is it possible that, with any aura of remoteness and inaccessibility quite expunged from far geographical locations, they no longer have the power to grasp our attention?

If this analysis of the malaise is at all to be believed, then perhaps an appropriate antidote is a trip around geographical locales that are intrinsically places of awe or puzzlement. The regions to which we make journeys of the mind should, perhaps, be those that still have the power to give scientific, rationalist humankind pause for thought and reflection. Let us travel the Great Rift Valley from end to end and ponder the forces that can make a rent thousands of miles long across continents. Let us pause in fascinated horror beside the seductively beautiful crater lakes of the highlands of the Cameroons—lakes that have killed thousands of their shoreline inhabitants in seconds. Let us go to the places of geographical mystery.

The Earth
riven apart

From space, the Earth looks serene and static, the shapes and positions of its continents established and well rooted. But, in fact, the Earth's crust is being torn apart along the longest continuous land crack on the planet—the Great Rift Valley.

This giant scar is a truly amazing and mysterious feature of planet Earth. It contains the lowest place on land, is flanked by some of the world's highest and biggest volcanic mountains, holds some of the world's largest lakes and provides a seaway between Europe and the Orient. From its northern extremity in Turkey to its southernmost expression in Mozambique, it measures 6,700km/4,200mls, almost one-fifth of the Great Circle that describes the circumference of the globe.

The rift begins in southern Turkey and runs south through Syria to Israel and Jordan, where the Dead Sea lies: the lowest point on all the continents. The lake surface stands at 400m/1,312ft below sea level, but its bottom is at 800m/2,625ft below the Mediterranean Sea, which is only 75km/45mls distant. The Rift Valley acts like an elongated basin, into which water flows, but out of which it rarely drains. Instead, in the high temperatures experienced in these regions, the water evaporates, making many of the rift lakes extremely salty. The salt content of the Dead Sea, for example, is 30 percent, 10 times higher than that of ocean water, giving it a buoyancy that allows a person to "sit" in the water.

From here, the rift passes south to Elat, some 800km/500mls from its starting place. At this point, the rift has been invaded by the ocean and runs the full length of the Gulf of Aqaba and the Red Sea, where it swings abruptly eastward between the Arabian Peninsula and the Horn of Africa. Now 3,000km/1,900mls long, the rift divides, one

One of the strangest, most uncanny sights on Earth must be Lake Natron, its vast expanse of shallow water stained red with a seasonal bloom of algae. Vast spirals of sodium carbonate, brought up from the depths of the Earth by geysers, decorate the surface, while slabs of soda rim the lake and float on its waters like miniature icebergs.

Only flamingoes can survive in this blistering environment, wading in the alkaline water and sieving it with their heavy beaks for the teeming algae that constitute their diet.

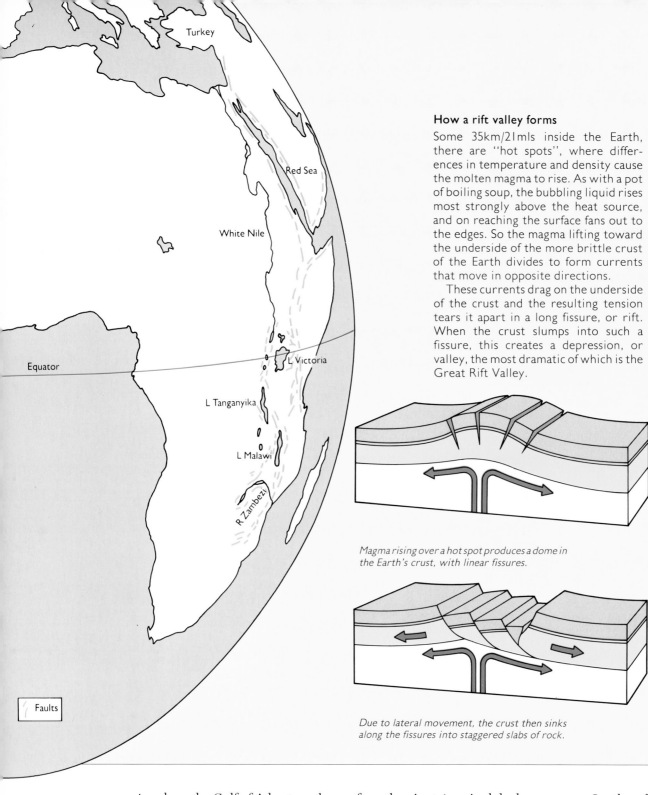

Turkey

Red Sea

White Nile

Equator

L Victoria

L Tanganyika

L Malawi

R Zambezi

Faults

How a rift valley forms

Some 35km/21mls inside the Earth, there are "hot spots", where differences in temperature and density cause the molten magma to rise. As with a pot of boiling soup, the bubbling liquid rises most strongly above the heat source, and on reaching the surface fans out to the edges. So the magma lifting toward the underside of the more brittle crust of the Earth divides to form currents that move in opposite directions.

These currents drag on the underside of the crust and the resulting tension tears it apart in a long fissure, or rift. When the crust slumps into such a fissure, this creates a depression, or valley, the most dramatic of which is the Great Rift Valley.

Magma rising over a hot spot produces a dome in the Earth's crust, with linear fissures.

Due to lateral movement, the crust then sinks along the fissures into staggered slabs of rock.

prong running along the Gulf of Aden to the Indian Ocean, and the other branch cutting inland into the Afar Triangle. One of the Earth's most inhospitable and geologically active regions, this area contains the lowest point in Africa, the Danakil Depression.

The rift continues down through the Galla lakes of Ethiopia—Ziwai, Shala, Abaya and Chamo—to Chew Bahir, the northerly limit of an expedition in 1888 by the Austro-Hungarian explorer Count Samuel Teliki (1845–1916). The count named the lake that then stood

there after the Austrian Archduchess Stefanie. But the climatic changes that were already in progress when Teliki chanced upon the lake have long since dried it up, so that today it is a salt flat (which is literally what Chew Bahir means) bordered by mysterious high mountains.

Passing south-southwest into northern Kenya, the rift runs through an area called Kino Sogo, where it is tearing apart the lava sheets that volcanic activity has spread over the landscape at various times during the last 15 million

years. South and west of Kino Sogo lies Lake Turkana. Previously called Lake Rudolf by Count Teliki in deference to his prince, it was renamed by Jomo Kenyatta, the first President of Kenya, for the fierce people who inhabit this corner of the country.

But Lake Turkana also has a nickname: the Jade Sea. If you approach it from the south, as Teliki did, it does indeed seem to have the opaque green coloration of that hard, smooth stone, due to the large quantities of algae that float in the water.

Near Lake Assal in the Danakil Depression, at 155m/510ft below sea level the lowest point in Africa, it is hot and dry beyond belief. The landscape is ancient, with deeply eroded rocks and heavy salt deposits. Yet it is ever changing, for it has experienced more recent volcanic activity than any other place in the Rift Valley.

Crystals of sulphur, brought to the surface by past volcanic activity, give an unlikely touch of golden yellow to this strange, stark desert landscape.

The bottom of the world

Although now separated from the Red Sea by a barrier of land, thrown up by disturbances in the oceanic crust, the fan-shaped Danakil Depression in northeastern Ethiopia has not always been dry land. Periodically the ocean has flooded this area, much of which lies 120m/400ft below sea level.

Evaporation of the sea water in the intense heat has left behind barren flats, covered with thick deposits of rock salt. In such infertile terrain, this is a precious resource, and the salt is mined in conditions of unimaginable heat and discomfort.

It is then transported by camel to the Ethiopian highlands, for sale in the markets of Addis Ababa.

Now the rift presses farther south into another of the Earth's "hell holes", the Suguta Valley, probably one of East Africa's least accessible places. The valley was once part of Lake Turkana, but volcanic eruptions created The Barrier—deposits of basaltic rock—which cut it off from the lake. Today the valley contains a remnant lake, Lake Logipi—but constant evaporation ensures that this is small and highly saline. No one lives in the Suguta, although it does from time to time form a temporary refuge for bandits.

From the Suguta Valley, the rift cuts toward Lakes Baringo and Bogoria. The climate here is less arid than farther north, although only modest amounts of rain fall throughout the rift in Kenya. Because of this limited rainfall, the natural vegetation of the region consists largely of thorn bushes, acacias and various types of grass, which although not deep rooted can tolerate extremely dry conditions.

By this point, the rift has again divided. The western branch contains Lakes Mobutu (formerly Albert) and Edward, and because the altitude is greater here, there is more rain, which in turn encourages the lusher vegetation of the rain forest. This branch then swings south, but is less clearly defined until it reaches the gigantic elongated lake of Tanganyika.

The main Rift Valley, the eastern branch, is known also as the Gregory Rift for John Gregory (1864–1932) the nineteenth-century physiographer and geologist. It passes west of Nairobi, flanked by the Aberdare Mountains and the Ngong Hills, the scene of so much strife during the Mau Mau struggle for Kenyan independence in the 1950s.

Two beautiful but strange lakes—Magadi and Natron—lie along this branch in southern Kenya and northern Tanzania. The weathering of local volcanic rocks has allowed rivers to carry large amounts of dissolved salts into these lakes, where high evaporation, due to the intense sunshine, has led them to crystallize. The salts are sodium bicarbonate, or soda, and in the dry season the glare off the white, rucked-up surface of salt is dazzling. After the rains, there is a thin veneer of water over the lakes, through which the soda pokes, looking like rafts adrift on a great sea. Only green algae can live in these alkaline waters.

The two East African branches of the rift meet again at the northern end of Lake Malawi (formerly Nyasa), which occupies a 580km/360ml long portion of the valley. But before they join, the two arms of the rift circumscribe Africa's largest lake, Victoria. A saucerlike basin with an area of around 69,490sq km/ 26,830sq mls, it is also the world's second largest freshwater lake. South of Lake Malawi, the rift opens out into the undulating plains of the Zomba Plateau and ends somewhat uncertainly in the swamps of the lower Zambezi River in Mozambique.

Because they occur above "hot spots" in the underlying magma deep within the Earth, rift valleys are often, but not always, accompanied by volcanic activity and eruptions. So it is with the Great Rift Valley. Several massive volcanoes dot the East African landscape. In Tanzania, the collapsed central part of the now-extinct volcano of Ngorongoro forms a wildlife sanctuary that is ecologically isolated from the surrounding grassy plains.

Farther north is Mount Kilimanjaro, Africa's greatest mountain at 5,895m/ 19,340ft high and 75km/47mls across at its base. North of Nairobi, in the same eastern branch of the rift, is Mount Kenya, which rises to 5,199m/17,058ft and is the second highest mountain in Africa. Although also a volcano, unlike Kilimanjaro it is now extinct. On the

edge of the western branch of the Great Rift is the non-volcanic Ruwenzori massif, with its many peaks over 4,877m/16,000ft.

Over its immense length, the Great Rift Valley touches the lives of 20 nation states. It contains some of the world's highest mountains and most profound depths; it produces some of the world's most spectacular scenery and a habitat for a huge part of its wild life.

Yet even more significant is the mysterious power with which the Great Rift Valley is forcing the continents of Africa and Asia apart. In Africa itself, the rending of the Earth's crust along the

Glistening, snow-topped Kilimanjaro, the greatest mountain in Africa, towers imposingly above the surrounding savanna. Visible from 160km/ 100mls in any direction, its cone shape reveals its volcanic nature.

The caldera of an extinct volcano, the Ngorongoro Crater lies some 208km/ 130mls west of Kilimanjaro. Now covered in rich, rolling grassland, it supports an amazing variety of game: zebra, wildebeest, buck and even a large population of black rhino.

fault lines of the rift may eventually allow the sea to flood in. Millions of years into the future, the coastal section of East Africa may partially split off from the rest of the landmass. In much the same way, that great island in the Indian Ocean, Madagascar, was created approximately 50 million years ago.

An ocean in embryo

Dividing a desert that stretches from the Atlantic shores of Mauritania, in west Africa, to the Gobi of central China, is a fingerlike stretch of water—the Red Sea. This impertinent intrusion, which almost completely separates the giant continents of Asia and Africa, looks insignificant on a map. But it is in fact the precursor of an ocean that might in time rival the Atlantic and which will grow at the expense of the present oceans.

Even at close range this sea with mysterious potential looks unimportant. On its eastern side is the aptly named Empty Quarter of the Arabian Peninsula, and on the western side the sparsely populated Eastern and Nubian Deserts of Egypt and Sudan. As a waterway it is 2,000km/1,250mls long and 260km/160mls wide. Yet it has few ports, and the land that fringes it has some of the lowest population densities in the world. Truly, the Red Sea's interest lies beneath its surface.

About 23 million years ago the Red Sea began to form as part of the rift, or tear, in the Earth's crust that stretches from Turkey, down through East Africa, to Mozambique. Where the African and Arabian continental plates have separated, the crust has collapsed between the plates and, gradually, over the millennia, seawater has invaded the resulting valley.

The Red Sea's development is a continuing process. Its more or less straight sides are moving apart at a rate of $1cm/\frac{1}{2}$in a year and have already separated by 320km/200mls at their widest point. It will be 100 years before the Red Sea is 1m/3ft wider, which may not seem impressive, but to achieve it, some 5×10^{17} tons of crustal matter will have been moved. If the continental plates continue to drift away from each other, the Red Sea will be flooded by the Indian Ocean, and it will grow from a "proto-ocean" to maturity.

A mysterious world apart, isolated from other oceans except for a narrow passage to the Indian Ocean, the Red Sea is ringed by blazing hot, sandy deserts. Its shores are fringed with corals that usually thrive only in the equatorial waters some 2,500km/1,550mls to the south, and its warm, salty waters teem with fantastic life forms, such as this angelfish, Pomacathus imperator, *its yellow stripes like sunbeams filtering through from the world above.*

As the Earth's crust is torn, magma rises from deep within it and is cooled rapidly by the sea to form a scab on the sea floor. Like any wound in skin, further tears in the same place are filled by more magma, which forms a new scab inside the old one. This process, known as "ocean-floor spreading", continues as the continents are forced farther apart. Although it is possible for ocean-floor spreading to slow down or even stop, there is every prospect that the Red Sea will develop further.

The lava that is building the new floor of the Sea brings with it fresh minerals that form deep within the Earth. These minerals, which include iron, manganese, zinc and copper, may yet prove to be its riches.

The fact that the Red Sea lies over one of the "hot spots" in the Earth's mantle has had several effects on the land. The principle one is the tilting away from its shores of the brittle continental plates that fringe the Sea. This has had two major impacts. First, there is no appreciable river drainage into the Red Sea, or into the Gulf of Aden for that matter. Second, the elevation of the Arabian Hijaz Azir plateau to heights up to 1,800m/6,000ft allows it to extract a little moisture from the Trade winds that originate in Asia and are usually niggardly with rain. This allows at least some agriculture in a peninsula which is renowned for its barrenness.

The high rate of evaporation in the area (the Sea has a potential evaporation loss of about 350cm/140in a year) and the lack of perennial rivers discharging into the Sea could spell its death if it were not for the water that flows into it from the Indian Ocean. The inflow through the Straits of Bab al Mandab, the neck of the Gulf of Aden, is phenomenal: its average rate is 0.25 metres per second, or half a knot—0.56 miles per hour—sufficient to compensate for the evaporation.

Another effect of the high evaporation and insignificant river inflow is that the salinity of the Red Sea is much higher than usual in ocean water. At the Straits, its maximum is 37 parts per thousand (3.7 percent) and this increases northward, until it reaches 40 parts per thousand (4 percent) off the southern tip of Sinai.

The dependence of the Red Sea on the Indian Ocean is absolute but tenuous. The maximum water depth at the lip just north of the Straits is a mere 159m/522ft, and the connection has not always been there. At times in the past, the Red Sea has instead received water from the Mediterranean, through the passage that is now the Suez Canal. This passage opened and closed several times during the Miocene epoch some 15 million years ago, due to the jostling of Africa against the smaller crustal plate of the Sinai Peninsula.

The rise and fall of the sea level during the last two million years, as the ice caps have grown and shrunk and ocean temperatures have fluctuated, has left its mark along the Red Sea. On the rocky slopes above the shore are fully developed coral reefs, with their beautiful symmetrical forms. These inanimate legacies of the Pleistocene epoch have been perfectly preserved by the arid environment.

It is ironical that so ancient a waterway, with such an inheritance from the past, should, in fact, seem set to become the great ocean of the future.

The formation of the Red Sea

The opening up of the seaway of the Red Sea has been a complex, multi-stage process. Up to 40 million years ago, Africa and Arabia were joined together. Between 40 and 15 million years ago (*top figure*) early continental rift formation created a discontinuity that ran through the present Gulf of Suez in the north, down the Red Sea Rift and then along the East African Rift.

Since 15 million years ago, this pattern has been altered by a combination of new rifting and ocean-floor spreading (*lower figure*). As Africa and Arabia moved apart from one another, the Gulf of Suez Rift and the East African Rift seem to have stalled. The Red Sea is, thus, now wider at its southern than at its northern end, and it opens into the Indian Ocean through the narrow Straits of Bab al Mandab and the Gulf of Aden.

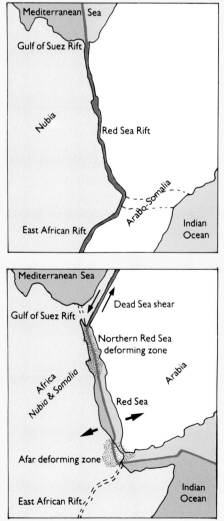

The rugged, arid coastline of Sinai's southern tip belies the rich life in the sea that laps around it. The currents in the Straits of Bab al Mandab tend to suck life into the Red Sea, rather than out, so enabling many unique subspecies of sea creatures to evolve there. For instance, there are more than 15 types of butterfly fish, Chaetodontidae spp, found only in these waters.

The Exodus enigma

One of the great puzzles connected with the Red Sea is how and where the Israelites succeeded in crossing it in their flight from the oppressive régime of the Egyptian pharoahs in about 1220BC. The first clue lies in the Red Sea's name, for this is a mistranslation of the Hebrew *yam suph*, meaning "Reed Sea", which probably referred to the marshy region just south of the Bitter Lakes. It is likely that the Israelites crossed here, since there are several fords, and the pursuing Egyptians on horseback and in chariots would certainly have become stuck in the thick mud.

It is also possible that the crossing was made at the head of the Gulf of Suez, where just such a strong wind as that with which God drove back the waters in the biblical story does occasionally blow, driving the sea back so far that people can wade across.

Mountains of the Moon

Early European explorers of the African continent stared in disbelief at their first sight of snow-capped mountains on the Equator. But the mountains were there all right: volcanic masses such as Mount Elgon (4,325m/14,190ft) and Kilimanjaro (5,895m/19,340ft), as well as non-volcanic mountains, among them the Ruwenzori, the fabled "Mountains of the Moon".

The great Anglo-American explorer Henry Stanley camped at the foot of the range in 1875, but its snows and glaciers remained hidden by heavy clouds, just as they still are for almost 300 days of the year. He only saw the glistening mountain tops—covered in "salt", as the local tribes-men believed—in 1888. Stanley called them the Ruwenzori, a corruption of the native *Ruwenjara*, meaning the rainy mountain. It is an apt name, for as much as 508cm/200in of rain, hail, sleet and snow falls there every year.

The story of the "Mountains of the Moon" is closely linked with the quest for the source of the Nile. The Greek historian Herodotus, visiting Egypt in 547BC, was told that the river rose from a bottomless lake between two sharp-pointed peaks. Then, in about AD50, Marinus of Tyre, a Syrian geographer, heard from a Greek merchant that, after journeying inland from the east coast of Africa for 25 days, he had reached two lakes and a range of snowy mountains, in which lay the source of the Nile.

In the second century AD, Ptolemy, the Graco-Egyptian geographer, summarized the existing information. He thought that the river originated from the union of two streams, flowing from two lakes far to the south, which were fed by melting snows on a range of mountains running east to west for almost 800km/500mls. These fabulous summits he named the *Luna Montes*, the Mountains of the Moon.

Remote and strange, the slopes of craggy Mount Baker loom out of cloud that hangs around the heights near Bujuku Lake like frozen breath. The local Bakonjo people believe that the mountain tops are the home of a great god, and amid their bizarre plants and ice formations explorers are filled with equal feelings of awe.

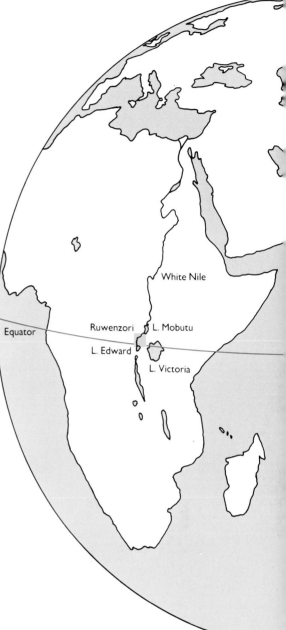

The Mountains of the Moon lie almost on the Equator, between Lakes Edward and Mobutu in the western arm of the Great Rift Valley. The only snow-covered range in Africa south of the Atlas Mountains, their torrential rains and gushing streams go to form the headwaters of the Mountain Nile.

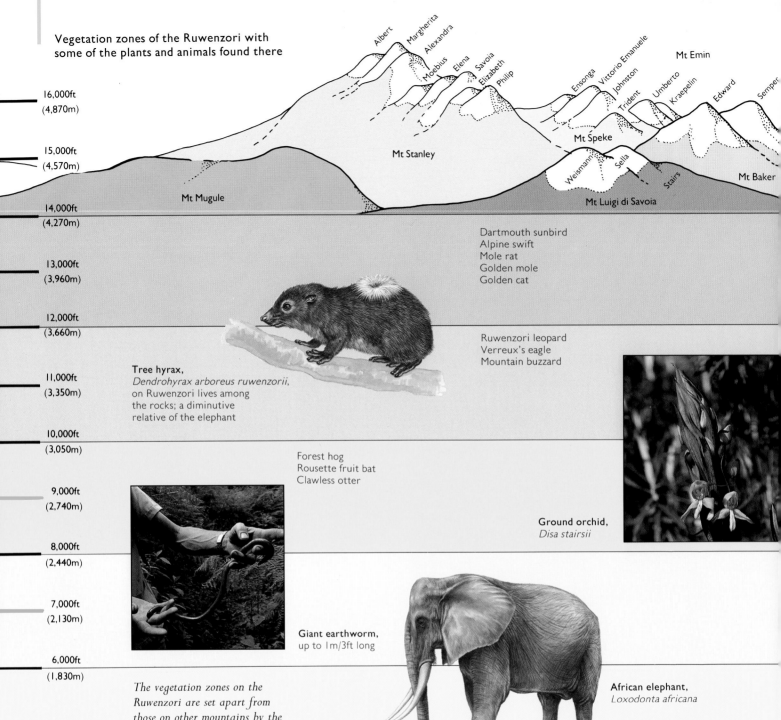

Vegetation zones of the Ruwenzori with some of the plants and animals found there

Albert
Margherita
Alexandra
Moebius
Elena
Savoia
Elizabeth
Philip

Mt Stanley

Ensonga
Victorio Emanuele
Johnston
Trident
Umberto
Kraepelin
Edward
Semper

Mt Emin

Mt Speke

Weismann
Sella
Stairs

Mt Baker

Mt Mugule

Mt Luigi di Savoia

16,000ft (4,870m)

15,000ft (4,570m)

14,000ft (4,270m)

13,000ft (3,960m)

Dartmouth sunbird
Alpine swift
Mole rat
Golden mole
Golden cat

12,000ft (3,660m)

Ruwenzori leopard
Verreux's eagle
Mountain buzzard

11,000ft (3,350m)

Tree hyrax,
Dendrohyrax arboreus ruwenzorii,
on Ruwenzori lives among
the rocks; a diminutive
relative of the elephant

10,000ft (3,050m)

Forest hog
Rousette fruit bat
Clawless otter

9,000ft (2,740m)

Ground orchid,
Disa stairsii

8,000ft (2,440m)

7,000ft (2,130m)

Giant earthworm,
up to 1m/3ft long

6,000ft (1,830m)

*The vegetation zones on the
Ruwenzori are set apart from
those on other mountains by the
strangeness of the life forms. Both
giant plants and unusual animals
have evolved here.*

African elephant,
Loxodonta africana

Although he placed the sources of the Nile too far south, Ptolemy's guess was not all that far out. His two lakes can be identified with Lakes Victoria and Mobutu (Albert), and his Mountains of the Moon approximately coincide with the Ruwenzori.

The Ruwenzori is one of the most inaccessible places on earth. It lies on the border of Uganda and Zaire, some 48km/30mls north of the Equator in central East Africa, between Lakes Edward and Mobutu. Although it is usually described as a mountain range, it is essentially a single enormous massif some 120km/75mls long and 48km/30mls wide at its maximum, rising to a group of four summits—Mounts Speke, Stanley, Baker and Luigi di Savoia— Mounts Emin and Gessi lying a little to the north.

The highest of the peaks, Margherita, reaches 5,109m/16,762ft, and there are eight others over 4,877m/16,000ft. The summits are separated by ravines hundreds of feet deep, cut by rivers that surge through them as raging torrents. The main stream, the Mobuku River, rises in the south, near Mount Luigi di Savoia, and flows east; it is joined by the Bujuku River from the northwest. The diameter of the snow-covered area is now no greater than 16km/10mls, although formerly it was much more extensive.

The massif owes its origin to Earth

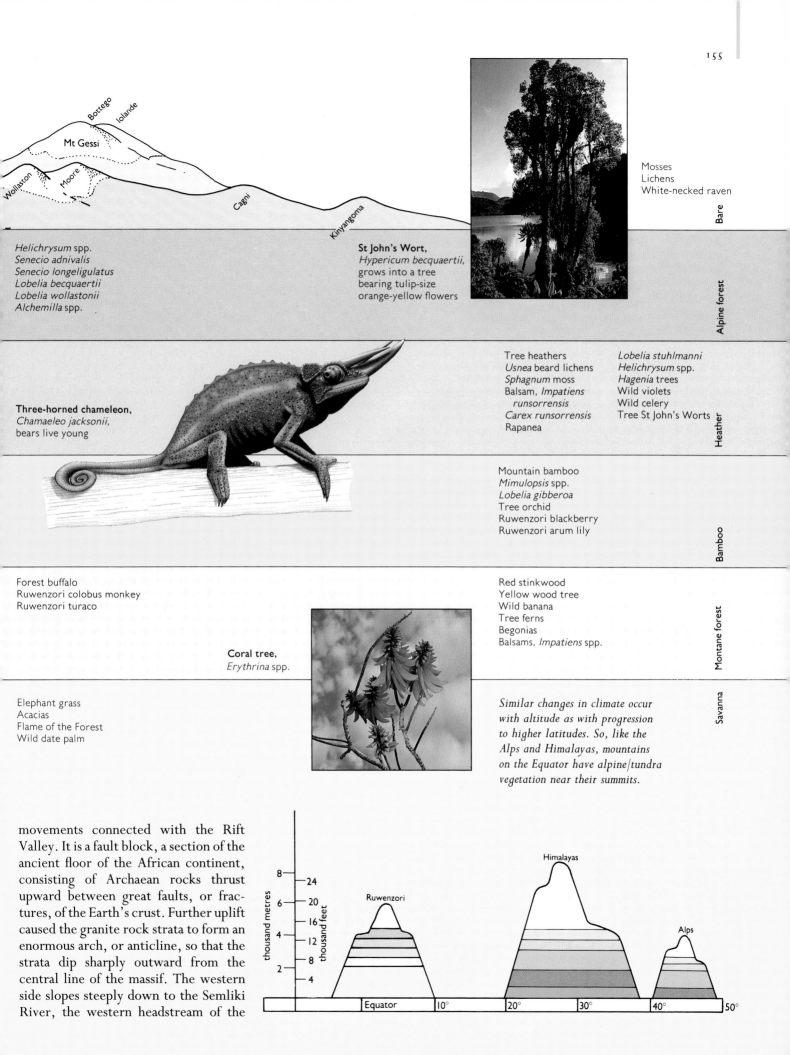

Mosses
Lichens
White-necked raven

Bare

Mt Gessi
Bottego
Iolande
Wollaston
Moore
Cagni
Kinyangoma

Helichrysum spp.
Senecio adnivalis
Senecio longeligulatus
Lobelia becquaertii
Lobelia wollastonii
Alchemilla spp.

St John's Wort,
Hypericum becquaertii,
grows into a tree
bearing tulip-size
orange-yellow flowers

Alpine forest

Three-horned chameleon,
Chamaeleo jacksonii,
bears live young

Tree heathers
Usnea beard lichens
Sphagnum moss
Balsam, *Impatiens runsorrensis*
Carex runsorrensis
Rapanea

Lobelia stuhlmanni
Helichrysum spp.
Hagenia trees
Wild violets
Wild celery
Tree St John's Worts

Heather

Mountain bamboo
Mimulopsis spp.
Lobelia gibberoa
Tree orchid
Ruwenzori blackberry
Ruwenzori arum lily

Bamboo

Forest buffalo
Ruwenzori colobus monkey
Ruwenzori turaco

Red stinkwood
Yellow wood tree
Wild banana
Tree ferns
Begonias
Balsams, *Impatiens* spp.

Coral tree,
Erythrina spp.

Montane forest

Elephant grass
Acacias
Flame of the Forest
Wild date palm

*Similar changes in climate occur
with altitude as with progression
to higher latitudes. So, like the
Alps and Himalayas, mountains
on the Equator have alpine/tundra
vegetation near their summits.*

Savanna

movements connected with the Rift
Valley. It is a fault block, a section of the
ancient floor of the African continent,
consisting of Archaean rocks thrust
upward between great faults, or fractures, of the Earth's crust. Further uplift
caused the granite rock strata to form an
enormous arch, or anticline, so that the
strata dip sharply outward from the
central line of the massif. The western
side slopes steeply down to the Semliki
River, the western headstream of the

thousand metres
thousand feet

Himalayas
Ruwenzori
Alps

Equator   10°   20°   30°   40°   50°

Nile, in the middle section of the great Rift Valley. The eastern side slopes more gradually to merge with the uplands of Uganda.

Ruwenzori is unique; not only is it the sole series of snow-covered peaks in Africa south of the High Atlas Mountains, but it possesses a most peculiar flora on its slopes and a strange fauna at its foot. This is partly a consequence of the ice age, for although the present snow-line is approximately 4,100m/13,450ft above sea level, the glaciers on its eastern side formerly reached down as low as 1,400m/4,600ft. During the last ice age, much of Africa was extremely cold, and extensive central and eastern areas were covered with giant flowering groundsels and enormous lobelias, with yellow wood trees, *Podocarpus*—a conifer with large leaves—and plants such as lady's mantle, *Alchemilla*. Today these huge forms survive only on mountain slopes covered in frost and shrouded in rain or mist.

As a result, at between 2,440m/8,000ft and 4,570m/15,000ft on the slopes of Ruwenzori, there is an "enchanted forest" of weird and grotesque mammoth plants, which wherever else they grow are small and insignificant. The open glades bristle with the upright stalks of lobelias such as *Lobelia woolastonii* and *L. bequaertii*, like green obelisks, growing 3.7m/12ft high.

These are mixed with the groundsels (*Senecio* spp.), whose writhing stems, crowned by heads of spiky leaves, appear like witches' brooms, and acres of everlastings (*Helichrysum* spp.), forming sheets of gold, pink or silver blossoms. And beyond them, on the precipitous slopes and ledges above, are tree heathers (*Erica arborea*) up to 12m/39ft tall, with their branches and trunks draped in swathes of yellowish moss and lichen which thrive in the conditions of constant moisture.

Many botanists think that the gigantic size of most of the vegetation is due to the intense ultraviolet radiation that comes from the Sun at high altitudes in equatorial regions, coupled with the mineral-rich and acid soil. But gigantism is a difficult biological phenomenon to

Early explorers

The first person to explore in the Ruwenzori was Stanley's assistant, Lt Stairs. From 1889 to 1905, the mountains were partially climbed by a succession of expeditions. In 1891, the German Dr F. Stuhlmann reached a point just short of the snow-line; it was he who realized that Ruwenzori consisted of several mountains and who detailed the distinct vegetation zones. The naturalist G.F. Scott Eliott, in 1895, discovered the route from the east up the Mobuku Valley that is now generally followed.

In 1900–01, five expeditions were made,

weather, a well-equipped expedition led by the Duke of the Abruzzi, a famous Italian sportsman and mountaineer, ascended all the principal snow-covered peaks in only six weeks. This expedition laid the foundation of all present knowledge and produced a 1:50,000 scale map which is of value even today.

Since local tribesmen had no names for the summits, the duke named them after early explorers, and the peaks after British and Italian royalty, previous climbers and scientists, and members of his own party.

This portrait of Prince Luigi di Savoia, Duke of the Abruzzi, was taken by the Italian photographer Vittorio Sella, who accompanied the duke to the Ruwenzori. Sella's remarkable pictures appear in Filippo de Filippi's 1908 account of the expedition, Ruwenzori.

Ruwenzori's mountains are capped by weird ice forms, as here on Coronation Glacier, between the peaks of Elizabeth and Savoia. In the intense heat of the equatorial Sun, exposed ice ridges form huge, rounded, overhanging cornices, often festooned with giant icicles that screen deep ice caves.

Expeditions usually set off from the Ugandan side and climb up to Bujuku Lake at 3,900m/12,800ft—the route that leads into the inner heart of the massif. From here, explorers and porters, right, must struggle through a pathless jungle of helichrysum and giant groundsel until they finally emerge on the snowy summits.

each one bringing back only very little further information about the mountains and their strange flora. And in 1905 the first attempt by a group of mountaineers to conquer the peaks was completely foiled by the atrocious weather.

But in 1906, in a remarkable spell of fine

explain, and some ecologists attribute the success of the monster species on the mountains also to lack of competition from trees, whose niche they have thus taken over.

The mountain tops of Ruwenzori, above the forest of heather, are not sharp or jagged but are, rather, domelike protuberances topped by small ice caps of the type usually known as secondary glaciers in the Alps. From the ridges of the loftiest peaks, particularly Mount Stanley, project broad cornices, like eaves, which are supported by ice-crystal columns. Fantastic curtains of pendant icicles, 15m/50ft long, that sway in the slightest breeze, tinkling like sleigh-bells, create countless ice grottoes on the steep slopes below the east ridge of Mount Margherita. These huge icicles form because, when the clouds disappear, the great intensity of the Sun's rays on the Equator quickly melts the snow and ice; but it freezes again equally rapidly at night, when the sky remains clear.

Ruwenzori is an appalling place weather-wise. A typical day starts relatively fair at dawn, and sunshine often lasts until 11am, when the mists appear and twist up the valleys, obscuring the summits around midday. All afternoon, until about 5 o'clock, heavy rain or snow falls from a dark and menacing sky, saturating everything. Then the Sun appears again, and the nights stay clear, cold and starlit.

But the weather can change its aspect in a fascinating and dramatic manner. During the rainy season, clouds from 160km/100mls around appear to be drawn toward the massif; they then steady for a time before being dragged down into its heart. Electrical effects follow each other incessantly, striking deep into the mountains. Thunderclaps of tremendous proportions occur within half an hour or so and then the rain begins, falling with incredible intensity. On other days, the mountains may generate their own clouds, throwing them out along the slopes so that, after a storm, the Sun re-emerges spectacularly from behind four or five massive cloud formations of startlingly different hues.

For centuries the tribes living in Ruwenzori feared it. Its rocks and snows, its torrential rains and the force of its rivers, the thunder and lightning that tried to tear it apart were all awe-inspiring omens to them, and they would not venture beyond certain points because of their superstitious dread.

According to Roger Redfern, leader of the British Ruwenzori Expedition in 1961–62, there is nothing in the Himalayas to compare with it for hard and morale-cracking terrain. It provides the worst walking in the world, a mixture of dense Amazonian jungle and Alpine altitude, and even the modern well-equipped visitor must expect to be lashed by biting rain, impeded by the over-exuberant plant life, lost in billowing mists and trapped in fearful bogs. In addition to the depressing physical impact these have on explorers, they must cope with the feeling of lethargy induced by the altitude and the inescapable sense of menace conveyed by the unrelenting wildness of the mountain.

The leader of the British Museum scientific expedition in 1934–35,

Tree heathers—a mammoth version of the usual scrubby bush—draped with long, trailing beards of Usnea lichens form a grotesque forest at 10,000–12,000ft/ 3,050–3,660m. Beneath them is the almost impenetrable carpet of other outsize forms of normally tiny plants. This reinforces the impression that this strange place might well be a "land of Pterodactyls and prehistoric monsters."

Graceful "torches" of Lobelia species, covered in long, delicate silvery hairs, rise from a rosette of slender leaves. As the plant grows, these leaves curl inward at night to protect the central bud and prevent it from freezing. Great tussocks of sedge, Carex runsorrensis, indicate the boggy nature of the ground.

Patrick Synge, felt that Ruwenzori was "the only mountain that had something to say to the traveller." He sensed "a feeling of personality, aliveness, resident in the mountain, something part of it . . . Although the silence was intense, we never felt that the mountain was passive. It was awake and watched our every movement." Once the adventurer in Ruwenzori is gripped by the almost occult quality of "weirdness and mystery, allurement and stimulation" that emanates from the mountain, he is never free from its siren call.

A chain of jewel-bright waters

An aerial view of the grandest fracture on the Earth's surface—the Great Rift Valley—reveals some of the world's most astonishing and bizarre "water-scapes": giant brown and white spirals oozing into surrounding blackness; red-stained honeycombed lake beds; white slabs of soda floating in steam; and veils of pink flamingoes fluttering over a jade-green surface. These are the East African soda lakes, just one variety of the many lakes that dot the Rift Valley.

Truly spectacular, East Africa's lakes are also a vibrant affirmation of the power of evolution to populate habitats, no matter how demanding, with life forms that are supremely adapted to cope with the conditions there.

The chain of East African Rift lakes begins in the north, with the Bishoftu crater lakes near Addis Ababa in Ethiopia. Farther south the rift divides into western and eastern arms. The eastern branch cradles such lakes as Turkana, Nakuru, Naivasha and Natron, whereas the western arm contains Lakes Mobutu (Albert), Kivu, Tanganyika and Malawi. In a shallow depression between the two arms is the largest African lake, Lake Victoria. These lakes have formed in four different ways, but all have origins in the massive tectonic movements of the crust that created the rift and in the volcanic activity that has frequently accompanied them.

Some of the lakes lie within "grabens", linear sections of the Earth's crust which have sunk between the rock walls on either side. On a massive scale, a graben looks like a single, angular stone in a flagged floor which has slumped downward. Where such a steep-sided slot occurs, watercourses are usually deflected into it, to form an extremely deep lake with a sharply shelving shoreline. Lakes Tanganyika and Malawi are both classic graben lakes.

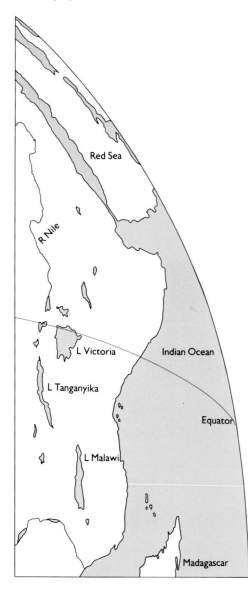

In the western branch of the Great Rift Valley are the deep lakes of Tanganyika and Malawi, while the eastern branch holds soda lakes such as Magadi and Natron. Lake Victoria lies in a shallow basin between the two major rifts.

In the waters of Lake Victoria—as great in extent as Switzerland—the mysterious forces of nature have, over the past million years, produced a kaleidoscope of cichlid fishes. Many are unique to the lake, and they provide a perfect example of the evolution of species.

Lake Tanganyika extends down to about 1500m/4,921ft at its deepest. In it, and in Lake Malawi, this great depth has a dramatic effect on water conditions. The waters stratify into layers of different temperatures and density. Because the surface waters, which are warmed by the hot equatorial African Sun, are less dense than the much cooler waters below, they consequently sit permanently above them. The bottom layers are incredibly dark and, because there is no real circulation of water between the top and bottom layers, they contain little oxygen. They therefore have no ordinary forms of life. The warm upper layers, however, vibrate with living things, and contain an almost unsurpassed number of lake fish species.

Lake Victoria, which is much shallower than the graben lakes (its greatest depth is only 80m/262ft), is a so-called tectonic lake. It was formed by crustal movements which produced an extensive basin into which the surrounding waters flowed. Its shallowness means that water mixes constantly at all the lake levels, so the lake bottom is oxygenated and receives some light. Consequently there is an abundance of food in the surface waters for creatures such as fish and birds, and also an extra set of niches for bottom-inhabiting animals.

The lake contains an extraordinary diversity of fish species (200 at least) many of them examples of the beautiful bright cichlids that typify these lakes. A single rock on the bottom of Lake Victoria can be home to several different cichlid species which are found only there—they are "endemic" to the lake. Some of these unique life forms have developed in the blink of an eye in geological and evolutionary time.

As an example, Lake Nabugabo, a tiny offshoot of Lake Victoria, was cut off from its parental waters only 3,700 years ago by the growth of an intervening sand bar. And in this brief time interval, the lake has already developed five unique species of cichlid fish different from any found in Lake Victoria.

Perhaps the best guess concerning the reason for the development of this

"species factory" is climate-linked. If the shallow lake bed of Victoria had dried up partially several times in the last few hundred thousand years, it would have turned from a single lake into a cluster of tiny separate ones. Successive periods of wet and dry climate in the Pleistocene Ice Age might have produced such habitat alterations. And on each occasion that this fragmentation occurred, it would have trapped small populations of fish which would, in time, have become specifically adapted to their own mini-lake. When the water level rose again, the different stocks would have been back in contact, but by then they might have diverged sufficiently to keep the lines genetically separate from one another.

The other types of rift lake are either volcanic or saline. In the volcanic type,

also known as soda lakes, volcanic geysers eject sodium carbonate, which gathers in spirals and slabs on the surface of the hot water. The Bishoftu crater lakes in Ethiopia, for example, have formed either in the circular craters of extinct or quiescent volcanoes, or at sections of a river where lava flows have blocked off the river's course.

Saline lakes result when the rainfall declines in the area around a lake which has no natural outlet; the lake shrinks and the concentration of mineral salts rises as the water evaporates in the strong sunshine. In some lakes, the final concentration of salts can become so high that most large forms of aquatic life find it difficult to survive. Instead, the lakes are inhabited by a mix of salt-tolerant life forms—bacteria, algae, minute crustaceans and worms. The

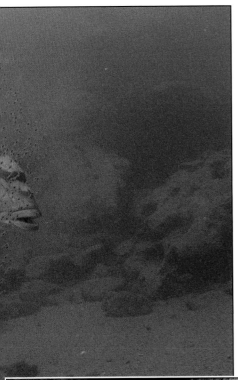

An astonishing range of body shapes, sizes and ways of life exists among the hundreds of cichlid species in the East African lakes. The omnivorous 38-cm/15-in Lobochilotes labiatus *fishes, with their cloud of fry, are one of Lake Tanganyika's 126 species, while the tiny hermit cichlid,* Pseudotropheus livingstonii, *popping out of a purloined snail shell, belongs to an incredible 200 species found in Lake Malawi.*

A voracious predator

The unique ecology of Lake Victoria is under increasing threat from the Nile perch, *Lates niloticus.* This predatory fish was introduced into the lake some 30 years ago in an attempt to provide another useful catch for the local fishermen. The piscine hunters found themselves in the world's second largest freshwater lake, filled with hundreds of species of small and medium-sized fish— a veritable paradise for a predatory fish. Few of the resident fish had any means to combat the formidable Nile perch, and almost all the populations of cichlids are now declining, while the numbers of Nile perch increase correspondingly.

Some scientists believe that general overfishing with fine-mesh nets is the main cause of cichlid losses in Lake Victoria. But there can be no doubt that the introduction of the Nile perch must have accelerated the demise of irreplaceable cichlid species.

The natural home of the Nile perch is in some head waters of the river, such as Lake Albert, and in the White Nile itself. Here it can grow to immense size, as this fish caught in southern Sudan shows. High waterfalls on the Victoria Nile, flowing out from the lake, prevented the fish from reaching Lake Victoria naturally.

bacteria and algae in some of these lakes endow them with the most extraordinary colours. In Lake Magadi, in Kenya, bacteria release iron oxide, which stains the bottom blood-red. When this lake dries, its carmine bed becomes spectacularly honeycombed with a hexagonal pattern of salt deposits.

The life forms that do, astonishingly, manage to survive in these mineral-laden waters, support a rich variety of larger animals. The saline lakes of Nakuru, Naivasha and Natron, for example, are famous for their flamingoes, whose remarkable beaks and tongues operate as sieves for straining microscopic food items from the soda-laden lake waters.

Nature's mysterious waterway

The Casiquiare waterway connects the two great river systems of South America—the Orinoco and the Rio Negro, itself a massive tributary of the Amazon. It appears to be a natural canal, the only one of its kind in the world. Yet the Casiquiare is more than a canal; it is a river in its own right, with a series of dangerous rapids and several large tributaries.

The Casiquiare is a unique geographical and hydrological fact, but its very existence has been the subject of passionate debate for hundreds of years, between explorers and scientists alike. It was initially reported in the mid-sixteenth century, at the time when conquistadores and missionaries first set out to capture the minds and lands of the native Indians.

Gold fever and tales of the fabulous El Dorado spurred them on. By 1641, a Padre Acuña had reported a waterway that linked the Orinoco with the Rio Negro to the south, but no one believed him. Exactly 100 years later, the Spanish Jesuit Joseph Gumilla published his book, *The Orinoco Illustrated*. His map showed the Orinoco rising in the Andes to the west and a great mountain chain separating that river from the Amazon, and he vehemently denied the existence of the connecting waterway.

Then, in 1744, another Jesuit, Father Superior Manuel Román, was guided down the Casiquiare to the Negro by Portuguese travellers, who were well accustomed to using the river as a slave route. The discovery was reported to the French Academy of Sciences in 1745. Father Gumilla was outraged and published a second edition of his book entitled *The Orinoco Illustrated—and Defended*. At least three more, well-documented Spanish expeditions navigated the Casiquiare over the next 20 years, but still scientists refused to believe in its existence.

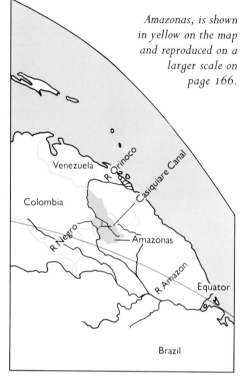

Amazonas, is shown in yellow on the map and reproduced on a larger scale on page 166.

"A stream without counterpart", whose very existence was for many years denied, the extraordinary Casiquiare, its waters yellowish-white with their burden of silt, meanders through miles of tropical jungle and links the Orinoco and the Negro/Amazon rivers.

Descendants of the warlike Guaica Indians, encountered by Humboldt near Esmeralda, use modern outboard motors, although they still pole their dugout canoes in shallow waters. In the nineteenth century, 25,000 people lived along the Casiquiare; today there are fewer than 50.

However, in 1800, the father of modern scientific geography, Alexander von Humboldt (1769–1859) set the matter to rest. He entered the Casiquiare from the Rio Negro and rowed upstream, taking 10 days to reach the Orinoco in an arduous journey against the rapid current.

Not only did he confirm the existence of the waterway, he also dispelled the strange notion that its flow was regularly reversed, depending on the water levels in the Orinoco or Negro at any one time. Humboldt wrote of his travels in 1816 in his *Voyage to the Equinoctial Regions of the New World*.

Several more recent scientific expeditions have gathered valuable data about the Casiquiare. In 1942–43, for example, a team of US engineers spent eight months studying the feasibility of getting wild rubber from the rain forests of the Amazon to the war factories in the USA via the Negro–Casiquiare–Orinoco route. The purpose of this circuitous inland journey, of more than 4,023km/2,500mls, was to avoid attack from U-boats in the Atlantic. In the event, the plan was never executed, but this and the 1968 British *Geographical Magazine*'s hovercraft expedition up the Casiquiare, which took five and a half hours hover-time to Humboldt's 10 days rowing, revealed interesting facts that may bear upon the waterway's origins.

The bed of the Upper Orinoco slopes gently and is covered with a thick, clayey deposit, brought down from the sandstone mountains, which has gradually raised it above the level of the surrounding countryside. When the river floods, water naturally flows to a lower level, and the Pamoni, the most northerly of the Casiquiare's tributaries, lies 3–4.5m/10–15ft below the Orinoco, about 32km/20mls away as the crow flies. The flood waters would, therefore, have had to flow only a short distance downhill to reach this existing river channel, so the Casiquiare connection may have been formed in this way over the centuries.

Other scientists think that the Casiquiare "captured" the Orinoco, hijacking some 25 percent of its head waters.

Strange waterways of Amazonas

Almost 460 years after Christopher Columbus set eyes on the delta of the Orinoco, the Franco-Venezuelan Expedition of 1957 discovered its source in the highlands of Guyana. Some 322km/200mls downstream, the Casiquiare links with it, then twists and turns through the dense rain forests of southern Venezuela for 354km/220mls. At its junction with the Orinoco, the Casiquiare is about 46m/150ft wide. It increases in width as it slopes gradually, in a gradient of 18cm:1km/11.7in:1ml, toward the Rio Negro, eventually widening to 457m/1,500ft—as "broad as the Rhine", as Humboldt exclaimed when he came upon it in 1800. The Casiquiare crosses the watershed separating the two great basins of the Orinoco and the Negro/Amazon at about 115m/377ft above sea level.

Why the Orinoco does not take this short-cut to the Amazon is a mystery. Instead, it continues its circuitous route of some 1,738km/1,080mls to the coast of Venezuela and the Caribbean.

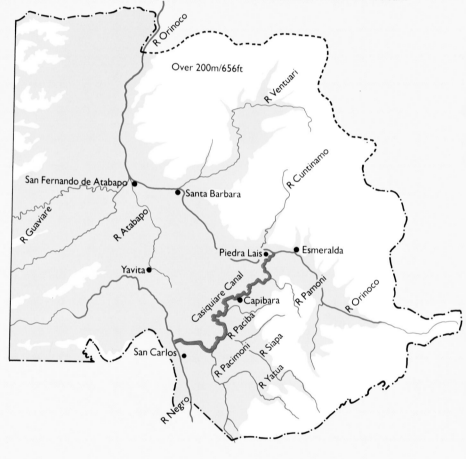

The two theories need not be mutually exclusive—the river-capture process could be complemented and augmented by the canalizing of the overflow waters.

A remarkable geological feature may also be relevant. Just before the junction of the Orinoco and the Casiquiare, there are two isolated cones of rock, the Cerro Tamatama on the north bank and the 27m/90ft high Cerro Doromoti on the south bank. Constricted, the river races between these obstacles and, once clear, turns sharply southwest. This violent current, over time, has eroded the south bank, helping to create the channel known today as the Casiquiare.

A curious phenomenon of the rivers of the area is the often startling contrast in the shade of their waters. Rivers such as the Orinoco that originate in the mountains have "white waters", which are turbid and loaded with eroded solids. As its name implies, the Negro, which rises in the lowlands, is a "black water" river. The darkness of its clear, acidic waters may be due to the rapid

Humboldt was intrigued by the outlandish animals he came across. This almost humorous painting of a brown howler monkey, now classified as Alouatta fusca, reveals him as an attentive observer of wildlife. Indeed, he is still regarded as the world authority on howler monkeys.

The tropical forest has not changed since this romanticized picture of Humboldt's camp by the Orinoco was drawn. Exotic flowers still hang from the trees and bright plumaged macaws, parrots and parakeets flash through the dark green foliage. Red howler monkeys defend their territories and raise the alarm when a prowling jaguar is seen. Herds of peccaries, or wild hogs, root for bulbs and tubers, and tapirs use their long mobile snouts to pick up leaves and fruit. Near the river live webfooted capybaras, the largest of the rodents, and in the water lurk caimans, piranha fish and dangerous coral snakes.

decay of organic material or, perhaps, to tannin in the water, derived from the trees and other plants, which also prevents insects from breeding.

This advantage is offset by the lack of nutrients and oxygen in black water rivers, and hence the lack of plant and animal life they will support. White water rivers, of which the Casiquiare is one for most of its length, are rich in nutrients, their banks are nourished by frequent floods, and crops such as cassava and manioc can be grown.

Undisputed masters of white water rivers are the insects—bees, ants, mosquitoes (zancudos locally), sandflies and tiny midges (jejenes)—that plague the flesh. Biting and stinging incessantly, they turn this beautiful natural wilderness into a "green hell".

The silent killer lakes

The grassy Cameroon highlands in West Africa are famed for some of the most stunning scenery in Africa. More than 30 large and majestic lakes, set in an undulating landscape of verdant valleys and mountains, stretch in a chain across most of the state. But the unruffled surface of these quiet pools holds a mysterious and sinister danger, which during the 1980s caused the deaths of nearly 2,000 people.

The lakes are crater lakes, visible reminders of the past activity of the volcanic Cameroon Line. This is a 700km/435ml long section of an even longer fault line, which extends as a series of islands of volcanic origin to Annobón Island, far out in the South Atlantic. Mount Cameroon, which erupted last in 1982, is the only currently active member of the chain.

The lakes are the main vents of ancient inactive volcanoes that have become blocked and have gradually filled up with water. Some of the lakes themselves are fated to disappear, as they slowly silt up and turn first into swamps and then into drier grassland.

When weathered and eroded, the volcanic rocks of the Cameroon highlands transform into fine, rich, mineral-packed soil. Ideal for some types of arable farming—cassava, yams and maize—this soil also provides extensive and nourishing pasturage for cattle. As a result, the highlands are dotted with farming communities of the Fulani and Bantu tribes. The Fulani are cattle herders, Muslims who moved into the region less than 100 years ago from areas farther to the north. The Bantu have lived around the lakes for thousands of years and are involved in mixed farming, ironworking and trade.

Their long association with the lakes has given the Bantu peoples a respect for them which sometimes amounts to worship. But this is no contemplative,

The scene looks idyllic: a calm blue lake set in a deep-rimmed saucer and fringed with lush vegetation; and beyond, rolling grassland and another, silted-up lake, whose fertile soil will soon produce rich crops for the local farmers.

But the beauty of these crater lakes of the Cameroons masks an unseen menace. Out of their dark waters, twice in the recent past, a cloud of poisonous gas has mysteriously bubbled up, without warning, bringing a terrible, silent death to all creatures for miles around.

calm reverence, for the stories of the lakes, handed down orally by these tribes, are shot through with a vein of unpredictability and, indeed, menace.

Some legends tell of lakes overflowing their banks and destroying villages. Other lakes are said to hold in their depths spirit-women who will pull fishermen down to their deaths; and stories abound of spirits killing all the fish and of lakes disappearing and then reappearing. To the casual outsider, visiting this marvellously scenic region, the threat implied by these legends, until recently seemed mere fantasy. But, in truth, the stories of the highland people must hark back to past disasters that have become part of folk memory, as the tragic events

that took place so dramatically during the 1980s suddenly revealed.

On 15 August 1984, the police in the town of Foumbot were notified that people were dying on the road near Lake Monoun. Together with a doctor, they went to the area and found bodies along the banks of the River Panke, which flows from the lake. They could see a whitish, smokelike cloud, as much as 3m/10ft deep, floating over the ground and drifting downwind off the lake. Since they were already feeling nauseated, dizzy and weak, they decided to go no closer.

It was eventually determined that 37 people near the lake had died mysteriously and abruptly that day. The event

occurred at a politically sensitive time and there were many rumours about its cause. Possibilities raised included the clandestine use of chemical weapons, the illegal dumping of chemicals in the lake, or some natural phenomenon linked with the volcanic nature of the lake. The Government of Cameroon and the Office of the United States Foreign Disaster Assistance immediately organized a team of scientists to search out the true cause of the disaster. Their findings were published in 1987.

The conclusion was that the disaster at Lake Monoun was, indeed, a natural one, caused by a lethal burst of gas from the depths of the lake. This was triggered by a landslide from the crater's

eastern rim, itself probably caused by a small earthquake.

Why, though, should gas emerge in huge quantities from a previously placid lake? Why should such gas be lethal? The answers to these questions lie in the volcanic origins of Lake Monoun. Although this volcano has not erupted lava for a very long time, it has continued to release volcanic gas, of which carbon dioxide is the main constituent, into the waters at the bottom of the lake. The immense pressure of the column of water above meant that large volumes of gas could be forced into solution in the water. It is just the same effect of high pressure that enables so much carbon dioxide gas to be held in solution in a pressurized bottle of fizzy lemonade.

The waters below 50m/164ft in Lake Monoun are completely devoid of oxygen and are saturated with dissolved carbon dioxide in the form of bicarbonate ions; they also contain large quantities of dissolved iron. Isotopic studies have revealed that much of the carbon dioxide infused into the water long ago, and they suggest that the lake is some 18,000 years old. It seems, then, that for thousands of years carbon dioxide had been seeping from underwater vents and increasing in concentration in the deep waters of the lake.

Most of gas remained in the dense, mineral-rich depths of the lake because of stratification—a layering of the water

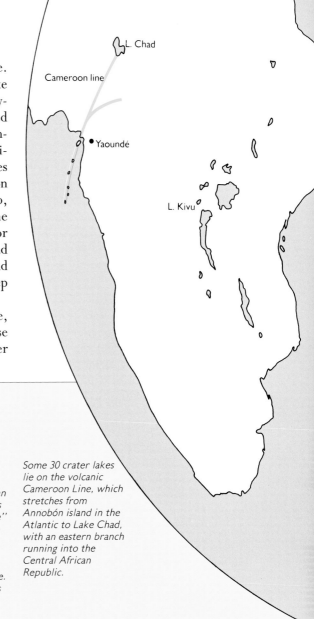

Gas burst at Lake Nyos

Gas kept in solution by weight of water

CO₂ seeps from underground vents

A landslip caused by an Earth tremor disrupts the well-layered "safe" lake conditions. Deep waters, with high concentrations of carbon dioxide, are brought to the surface. This death-dealing gas spurts from the lake and rolls down the valleys to bring destruction to nearby villages.

Some 30 crater lakes lie on the volcanic Cameroon Line, which stretches from Annobón island in the Atlantic to Lake Chad, with an eastern branch running into the Central African Republic.

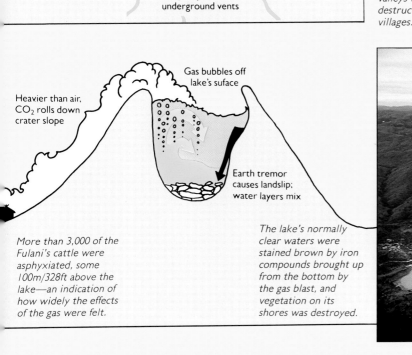

Gas bubbles off lake's suface

Heavier than air, CO₂ rolls down crater slope

Earth tremor causes landslip; water layers mix

More than 3,000 of the Fulani's cattle were asphyxiated, some 100m/328ft above the lake—an indication of how widely the effects of the gas were felt.

The lake's normally clear waters were stained brown by iron compounds brought up from the bottom by the gas blast, and vegetation on its shores was destroyed.

which means that surface and deep waters never become mixed. Over the centuries, small amounts of carbon dioxide must, however, have risen to the surface and passed imperceptibly into the atmosphere.

On that August day in 1984, this slow process was abruptly terminated. Around midnight on the night before the disaster, people near Lake Monoun reported hearing an explosion or loud noise; others remembered the disturbance as a slight earthquake. The tremor started a landslip on the crater rim of the lake, which dropped appreciable amounts of rock and soil into its depths. This could have been the fateful trigger that disrupted the neat stratification of the lake's waters: a violent stirring in the depths that caused "overturn"— the movement of deep waters to the surface.

The overturn produced a result rather like the uncorking of a bottle of champagne. Water containing huge amounts of gas, held in solution by pressure, moved up toward the surface, where the pressure was lower. As it rose, the carbon dioxide effervesced in a storm of bubbles, which themselves encouraged more to be produced. The gas burst was so huge and convulsive that it pushed a wave of water in front of it up the lake shore, flattening vegetation around the lake to a height of 5m/16ft.

But the wave did not kill; it was the gas that killed 37 people. Carbon dioxide is an asphyxiating gas; if one breathes it in in high concentrations, death ensues very rapidly. It is difficult even to hold one's breath, having inhaled one dangerous lungfull, because low oxygen and high carbon dioxide in the blood signal to the human brain that it is necessary to breathe more deeply. This reflex action simply hastens suffocation.

When the report of the scientists investigating the tragedy at Lake Monoun was published in 1987, it had, however, been overtaken by events. On a summer evening, 21 August 1986, a loud roaring noise, like that of a low-flying aircraft, was heard near Lake Nyos, another crater lake 120km/75mls north of Lake Monoun. Soon after, a

terrible column of vapour and gas rose from the lake and poured like a river down adjoining valleys. This cloud of destruction, 46m/150ft high, flowed 16km/10mls downhill. When it had passed, 1,200 people were dead in the village of Lower Nyos. In other nearby villages 500 more perished in minutes.

Apart from its horrifically greater death-roll of more than 1,700 people, the Nyos catastrophe is eerily similar to that at Monoun. Earthquake activity seems to have flipped a stratified safe lake into a turbulent, death-dealing one. There are real fears that the region of the Cameroon highlands is entering a period of increased seismic activity. This may increase the leakage of gas, trigger more lake overturns, and mean that lake after lake could do to its shoreline communities what Monoun and Nyos did within two years of each other. The lake legends of the highland people seem to have encapsulated accurately the malignant aspect of these beautiful waters.

Villagers, fearing the deadly and mysterious power of the crater lakes, make sacrifices to appease the spirits that live in their depths. Here they prepare a concoction of herbs and chicken's blood, which they will then pour into the water at a sacred spot in Lake Barombi Mbo.

Wearing masks, and flanked by dancers, the tribes' spiritual leaders held a great "cry-die" ceremony to mourn the victims of the catastrophe at Lake Nyos. In this way, they could assuage some of their anger at the noxious evil that had claimed so many lives.

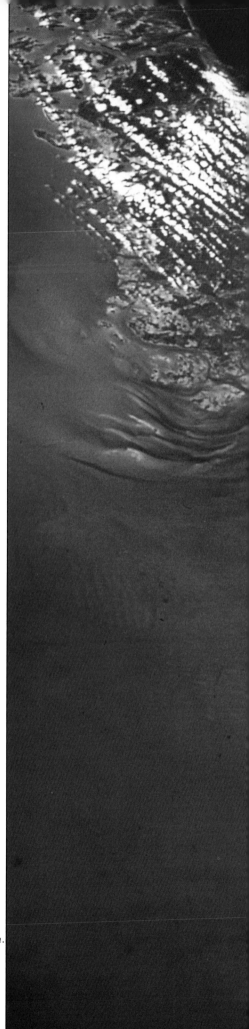

Hidden lairs of the Lusca

The turquoise, sandy shallows of the Bahamas are dappled with deep azure, like flicks of colour from an artist's brush. Many of these patches are the "blue holes" for which the region is renowned, and while to tourists they are merely confirmation of the beauty of the Caribbean, local fishermen treat the blue holes with circumspection.

Legend knows them as the home of the Lusca, a creature that is half-shark, half-octopus, whose indrawn breath causes the blue holes to become rolling whirlpools and whose grasping "arms" draw debris, men, and even boats down into the watery gloom. Later, a heaving, mushrooming dome of water is exhaled to the surface—the Lusca is satisfied.

These blue holes are actually the entrances to underwater caves found in considerable numbers on the margins of islands. Their equivalents on land are black circular lakes in the green, wooded parts of the islands.

The strong respect the locals have for the "boiling holes" is matched by that of the divers who are now exploring them. The hazards of these unique features, and the legends attached to them, have meant that for centuries they went unexplored and unexplained. Only in the 1950s were the first attempts made systematically to chart and investigate these timeless vaults. Exploration is now revealing fascinating information about the past history and modern geological processes that shape the Bahamas.

The Bahama Banks are unique in that they are the largest body of carbonate sediments that is actively building. The islands that rise above the sea are formed from layers of limestone beds between 3km/2mls and 6km/3.7mls deep. They occupy a fragment of the fringe of the North American continental plate that split away from the African plate when the two continents parted some 190 million years ago.

Since that time, lime-rich sediments, or carbonates, have been deposited almost continuously. Such sediments are laid down especially rapidly in the tropics, where warm waters and sunny

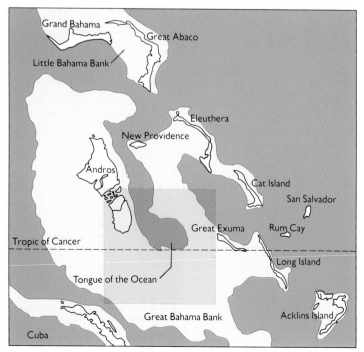

Andros Island, which sits on the marine plateau of the Great Bahama Bank, is lapped on three sides by warm, shallow seas. But in the east, a valley of water more than 1,800m/ 6,000ft deep—the dark-blue "Tongue of the Ocean"—curls past the shore. Along the wall of this abyss lie the fascinating and dangerous "blue holes" of the Caribbean.

The yellow square on the map covers the same area as the spaceshot opposite.

Inland holes act as windows on the freshwater lenses beneath the islands.

skies encourage evaporation and direct chemical precipitation. These conditions also permit the rapid growth of many organisms whose skeletons, which are composed of calcium carbonate, add to the layers forming the limestone. This "carbonate factory" is most productive in shallow banks, such as those surrounding the Bahamas, where the light penetrating to the sea-bed allows shallow-water organisms such as corals to flourish.

Corals need plenty of nutrient-rich water to survive, so they develop around the margins of islands and banks, forming a rocky barrier which encircles the island or bank. Just behind this barrier, coral debris accumulates; farther into the lagoon muddy sediments prevail, and in the shallows, the roots of mangroves trap and bind the sediments. The front of the reef is often very steep and is known to divers as the "wall" or the "drop-off". It results from natural cementation of the surface and interstices of the reef and from continued upward growth by massive corals. Water circulating within the wall may dissolve it in places, creating subterranean caverns.

Rainwater, which gathers on the land and in the rock, lies like a lens over the underlying saltwater, since saltwater is heavier than freshwater. Where the two meet, they form a powerful mix that dissolves the surrounding rock more quickly than either fresh or saline water

| Seawater | Freshwater | Brackish water |

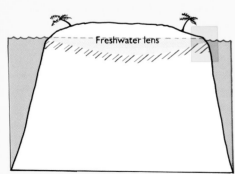

Rainwater, seeping through overlying rock, lies like a lens on the saltwater, which is heavier than freshwater. Where the two mix, the resulting brackish water dissolves the rock more quickly than either fresh or saline water.

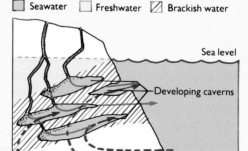

The tidal flow at the base of the lens removes the brackish water through fissures in the rock, eroding them. These channels eventually become caverns.

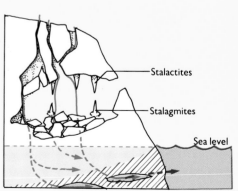

During successive ice ages, water was locked into ice caps and glaciers, and the sea level fell. Underwater caverns which had already formed were drained. Stalactites and stalagmites developed in the air-filled caverns and, no longer buoyed by water, many roofs collapsed.

does. When the tidal flow at the base of this lens removes the water through fissures in the rock, these channels are eroded, and eventually they, too, form caves.

During the last two million years, successive ice ages locked water into ice caps and glaciers and caused the sea level to fall. At the peak of the last ice age, 18,000–20,000 years ago, sea levels were as much as 120m/394ft lower than at present. This meant that the underwater caves that had already formed, were drained. No longer buoyed by water, their ceilings collapsed.

In many cave roofs a dome developed upward and in some places reached the surface, forming an open shaft. When sea levels rose again, the shafts filled with water and created the blue holes. In one cave on South Andros island, this rise innundated the nest of an extinct owl, which has recently been discovered by cave divers.

Stalagmites and stalacitites formed in the caves when they were filled with air. This formation stopped each time the sea level rose because the deposition could not continue, but it started again with the subsequent ice age when the sea level fell again. These formations can now provide scientists with excellent information on past sea levels.

At high tide, the level of the sea rises higher than the groundwater on land. The pressure of the sea therefore forces

Exploring underwater caves is perilous, so divers carry numerous lights to penetrate the eternal blackness, and double up on airtanks, regulators and even masks, for safety's sake.

water down into the blue holes, creating their strong characteristic whirlpools. Groundwater levels rise slightly at this time. At low tide, when the seawater pressure has dropped, the groundwater pushes the seawater down, causing it to well out of the mouths of the blue holes in great domes. It is these strong reversing currents that have given rise to the myth of the Lusca.

Divers take the utmost precaution when venturing into the blue holes.

Exploring underwater caves normally requires specialized techniques—and a cool head—because there is no ready escape to air or the surface. Movements in the extreme conditions of the blue holes have to be slow and considered because a careless stroke can stir up sediment and reduce visibility to zero. This is why cave divers always lay a guideline through the underground maze—an Ariadne's thread which leads them back to the entrance and safety.

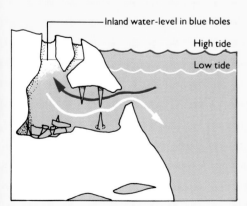

At high tide, sea level is higher than groundwater level, so the sea flows into the caves and groundwater level rises. At low tide, seawater pressure and, therefore, groundwater level drop. As a result, seawater floods out of the caverns.

The waters of the Black Hole on Andros Island appear inky because they are so deep. More than 1km/½ml in diameter, the hole was formed when the roof of a huge cavern collapsed and rising seas at the end of the last ice age filled it to the brim. Its crystal-clear depths have never been explored.

Worlds under the ground

Deep beneath the dense, tropical greenery and humid air of the Gunong Mulu National Park in Sarawak, Borneo, is a cavern so huge that eight jumbo jets could park nose to tail across the middle. Sarawak Chamber, as it is called, is 700m/2,297ft long by 400m/1,312ft wide: the world's largest known underground cavern. Amazingly, it is totally unsupported other than at its sides. This is because the limestone beds out of which it is carved are particularly thick and unbroken by joints.

Most of the world's caves, and certainly the largest of them, are found in limestone. Others, such as the extensive mazes of the Optimisticeskaja Cave in the Ukraine, which is 165km/103mls long, are in gypsum. Yet others, such as the enormous and ancient shafts of the Sarisarinama Plateau in Venezuela, have been sculpted out of quartzite rock.

In all rock types, caves develop when water and the material it carries dissolves and erodes the bedrock. Pure water has very little effect on limestone. But when water is charged with carbon dioxide it becomes acidic and can dissolve larger amounts of rock more rapidly. The air around us contains a small percentage (0.03%) of carbon dioxide, which is enough to make rainwater a very mild solution of carbonic acid. This water sinks into the limestone through fractures in the rock and erodes runnels and grooves. The dispersed rivulets eventually gather underground as a single flow, and the channel they carve becomes a cave passage. However, because much of the water's acidity is used up near the surface, the erosion process is extremely slow: the rock is eaten away, typically, by just an inch or so every thousand years.

But cave formation can take place more rapidly. For example, where rainwater is concentrated into rivers and

Dwarfed by the immense arch of the roof of Deer Cave in Sarawak, explorers appear totally insignificant. Even here at the cave entrance, small pools of light from the torches on their helmets seem hardly more than the fitful light of glow worms.

Humans are creatures of the light, and cave explorers can suffer the most primitive and deep-seated fears—of the enveloping darkness, of suffocation in narrow passages, of terror in great echoing chambers that seem to have no limits and no exit. And yet, drawn on by the mysterious tug of the unknown and the alien, cavers venture again and again to penetrate the labyrinthine worlds under the ground.

streams on impermeable rocks such as shales, which are adjacent to or overlie the limestone, caves are created at a faster rate because of increased exposure to the eroding acid. This was so for Deer Cave in Sarawak, the largest known cave passage, which measures more than 100m/328ft high and 150m/492ft wide.

In addition, the larger the flow of water, the more it can carry quartz sand, cobbles and, as in Sarawak, large boulders, which abrade the soft limestone and further augment the rate of cave development. In the tropics, erosion rates are very high compared to temperate areas because, with more abundant rainfall, the water is flushed continually—and at a faster rate—through the cave systems. So it is not surprising that the world's largest cave is in the tropics.

Cave streams often flow into surface rivers via springs or resurgences. As the river cuts its valley deeper into the underlying rock, so the cave streams dissolve new courses lower in the limestone to keep at the same level, and the initial cave passages, perched above the valley level, are abandoned. Much of the extensive network of passages in the Flint Ridge-Mammoth Cave System in Kentucky, at more than 500km/312mls the world's longest cave, are dry "fossil" passages of this type, abandoned by their formative streams as the Green River cut down its valley.

But abandoned passages are rarely completely dry. Water usually manages to seep down from the surface through tight fissures in the rock, and drips from the roof. This water has percolated through the soil where respiration by plant roots and the breakdown of organic matter by bacteria create concentrations of carbon dioxide 100 times higher than in the open air.

The water that soaks through the soil thus becomes even more acid than rainwater and is able to dissolve and contain much more limestone, in the

The persistent drip of calcium-rich water creates fantastic formations, such as these "organ pipes" and round-topped stalagmites in the Luray caves of West Virginia.

form of calcium carbonate, than surface runoff can. But when this water enters the well-ventilated cave, where there is little carbon dioxide in the air, the carbonate ions in the water are converted to carbon dioxide gas which diffuses out of the water into the cave air. The water is chemically unable to hold the calcium carbonate, or calcite, in solution and so it is deposited on the cave roof and floor.

These calcite deposits are responsible for some of the most amazing natural formations. Common examples of this cave sculpture are stalactites, where drips of calcium-containing water from the roof leave their mineral load behind to form columns suspended from the cave ceiling. Stalagmites form where the drip has been more rapid, so the water has dropped on to the cave floor, depositing calcite, which has developed into a rising column. In places, the two types of column meet to make a solid structure.

In cold climates, the low biological activity in the soils limits carbon dioxide production, which means that limestone is not eroded quickly and calcium carbonate levels in the solution are not very high—stalagmites and stalactites are therefore very sparse. However, in warm climates high concentrations of carbon dioxide develop in the soil, and cave formations are plentiful and spectacular. The rate of growth of these

speleothems, the scientific name for all cave formations, therefore, provides scientists with a unique record of past climatic variations.

Cave formations can also be dated, by a system known as uranium series disequilibrium dating. When calcite is

Delicate soda straws hang in profusion from the roof of a cave at Craig ar Ffynnon in Wales. Straw stalactites form when water dripping from the cave ceiling deposits calcite around the top of the drip, slowly forming a tube, below. If the tube becomes blocked, water runs down the outside, the deposit thickens, and a stalactite forms.

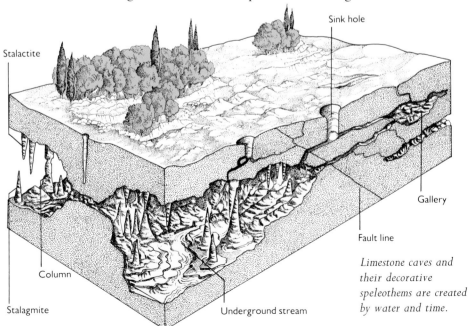

Stalactite

Sink hole

Gallery

Fault line

Column

Stalagmite

Underground stream

Limestone caves and their decorative speleothems are created by water and time.

The fascination and fear most people feel for the dense-black underground world is strong, and caves have long been a tourist attraction. This leisurely boat trip on the Echo River, in the Mammoth Cave System of Kentucky, was photographed in the mid-1890s.

pottery, which permit us to trace human development through time.

Anyone who has stood inside a vast cavern, or who has explored a labyrinth of winding underground passages, must be awed by caves. We can sense just such a feeling of wonderment and superstitious dread in the paintings of early cave dwellers. Imagine these men deep underground, with only the guttering flame from a reed torch or smoky-wicked oil lamp; out of the shadows spring a myriad of images: horses, bison, deer leaping above on the roof and racing along the walls into the shadows beyond. Even today, people who gaze upon cave paintings share this wonder and feel the power of this earliest art.

What inspired these men to venture into the dark, often for considerable distances underground? Perhaps the paintings depict game which the hunters hoped to slaughter in the chase; perhaps they were symbolic of the tribal family, or even, in the case of the stencilled hand prints, simply to express a sense of personal identity, to say "I was here." It seems more probable that they were of ritual significance, representing a spiritual world. But the answer is likely always to remain a mystery—the secret lost with the last of their race.

deposited it contains uranium. Over time, a form of uranium (the 234 uranium isotope) decays to become a form of thorium (the 230 thorium isotope). The ratio of the isotopes of these two minerals reveals how much time has passed since they were deposited. For example, using this method, scientists have been able to determine that, in Sarawak, the dry "fossil" caves only 30m/98ft above the present river valley floor are more than 350,000 years old. Taking this dating further, to the

highest levels of Deer Cave, suggests that it is nearly 2.5 million years old, which could help to explain its size.

Much of an original void may, in time, be filled in by speleothem deposition in caves abandoned by their formative streams. This may be augmented by the collapse of the roof and by sediments washed in from the surface. It is in these deposits, accumulated at the entrance to caves, that we find evidence of early human occupation: remains of bones and plants, discarded tools and

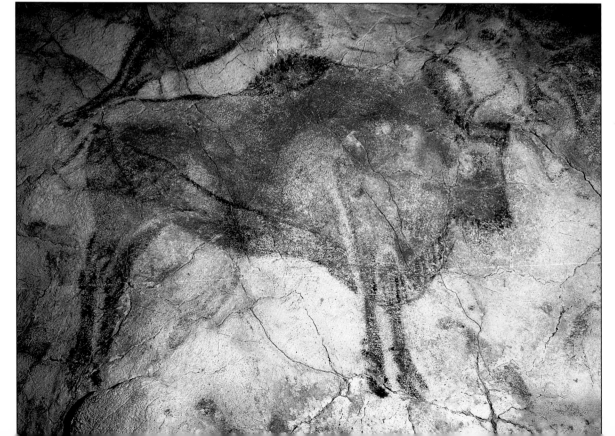

The paintings of animals that cover the walls and ceilings of caves at Altamira, in northern Spain, probably had a mystical significance for the artists.

Exploring caves is arduous and dangerous. The entrance pitch at Calf Holes, Yorkshire, is a vertical drop into the wet darkness, and the caver has urgent need of his strong, metal-runged ladder, securing rope and safety helmet. But his greatest security lies in good teamwork.

Living on the Earth

Of all the planets in the Solar System, Earth is, it seems, the only one that can support life. The fact of life on our planet is, in itself, deeply mysterious; it is no less awesome since biologists and climatologists have elucidated some of the factors that crucially determine its existence.

All life exists within the "biosphere"—a breathtakingly simple yet beautiful concept. This nearly spherical volume of space that encompasses the world is a shell of life only 27km/17mls thick, yet it contains more than a million species of living things. It descends some 11km/7mls into the profoundest ocean depths and stretches 16km/10mls or more up into the atmosphere, where living spores and pollen grains float high above the Earth.

Ever since the surface of our world—its crust—solidified more than four billion years ago, most of the planet's history has, in fact, been linked to the story of life on Earth. For life began soon after there were oceans, and evidence from tiny microfossils tells us that more than three billion years ago bacteria-sized creatures inhabited the seas. And it is plausible that even before that time there were living things in the oceans that were too minute and evanescent to have left behind any perceptible record in the rocks. At each stage in the Earth's history, from those early days onward, there has been a mysterious interplay in progress between the physical world and its living inhabitants, which has taken place on both an intimate and a global scale.

The very atmosphere of today's world has been shaped by individual living creatures. The primeval atmosphere contained no oxygen. That gas began to accumulate only after the advent of chlorophyll-containing organisms—the first of them the minute blue-green algae. By the process of photosynthesis these organisms trapped carbon dioxide and, as a result of their day-to-day biochemistry, expelled oxygen. Thus the oxygen in the modern biosphere has built up over the millennia as the algae, and their more complex green descendants, have produced immense volumes of this gas.

Animals, which depend on oxygen for life, could evolve only after sufficient levels of oxygen had built up in the atmosphere. The remnants of countless generations of our animal predecessors survive on Earth as a reminder of our mysterious evolutionary past. Many of the rocks beneath our feet are no less than the recycled skeletons of minute sea creatures. The gleaming white face of a chalk cliff represents trillions of tiny animal skeletons each, when alive, a single cell in size. The flints in that same cliff are silica, extracted by sponges from long-gone ocean water to construct their own body framework.

Coelacanths were known only from fossils until one was caught in 1938.

Indeed such a cliff is a microcosm of the inexorable flux between the planet and the life it has nurtured.

When life is viewed in this elongate and holistic perspective, the stories of today's living creatures and those of a thousand or even a hundred million years ago, meld into an unbroken chain of existence. We can learn much about the intricacies and mysteries of life by studying the creatures that are all about us today. But we can widen our knowledge far more by scrutinizing the fossilized remains of organisms long extinct. Suddenly the study of fossils is no longer the province of dusty old professors rooted in the past. Rather it becomes an amazing window through which to view the unfolding mysteries of the history of life on our planet.

Fossil perfection

Human beings see what they have been conditioned to see. Our attention is caught by what we know, and our interpretation of the world around us is influenced by experience. In no field is this more true than in the study of fossils, those mysteriously preserved remains of ancient life forms.

Today, most people know that fossils are the surviving parts of long-dead living things preserved within rock and as rock. Fossils are everywhere. Cliffs, road cuttings, seashore rocks, quarries and mountains are studded and encrusted with them. It is even possible to go on a fossil hunt in a city

Many of the grand buildings in the financial heart of London, for instance, are built of Portland or Purbeck limestone from southern England. Walking along a street, one can run one's fingers over the remains of shellfish, ammonites and corals of the Jurassic age. These fossils are known to be more than 150 million years old, and it is an extraordinary and almost unnerving bonus that we are able to look directly upon forms of life from a incomprehensibly remote period of time.

The very idea of the immense age of the Earth's rocks, and of the fossils within them, is a relatively recent one. For close on 1,800 years, European thought was dominated by Christian doctrines and beliefs, and a literal reading of the chronology implicit in the Book of Genesis in the Bible was generally accepted. This suggested that the world and its creatures had been created only in 4004 BC.

In such a climate of thought, fossils were regarded as either the artefacts of ancient peoples, debris from ''known'' biblical events such as the Flood, or miraculous products of more recent origin. In Robin Hood's Bay on the Yorkshire coast of England, for example, coiled Jurassic ammonites of the

The sheer abundance of ancient life forms, and the ingenious and elegant adaptations that enabled them to survive for millions of years, are strikingly evident on Charmouth Beach, on the Dorset coast of England. Embedded in the hard blue limestone (lias) cliffs, which time and the sea are steadily eroding, are the fossils of countless ammonites, whose only surviving shelled relative is the nautilus.

The coiled shell protected the ammonite and also kept it afloat, for the rear segments, added to as the animal grew, were hollow. They acted as buoyancy tanks, which could be flooded at will.

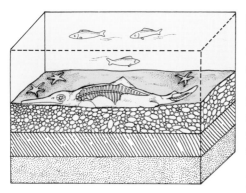

An ichthyosaur dies and falls to the seabed; flesh decays but bony parts remain.

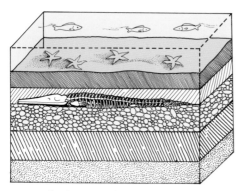

Layers of sediment settle on the bones and minerals from the sea slowly replace them.

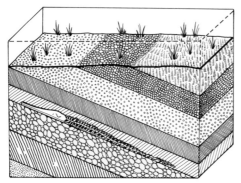

The fossil skeleton, now on dry land, is compressed and distorted by land movement.

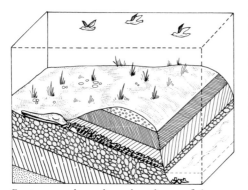

Erosion cuts down the rock and part of the fossil ichthyosaur is exposed.

genus *Dactyliocerus* were thought to have been dangerous snakes, helpfully turned to stone by the miraculous intervention of Saint Hilda. In medieval times, local stoneworkers often "improved" upon natural geology by carving snakes' heads and eyes on these ancient relatives of the squids. People saw what they wanted to see in the fossils.

The modern scientific consensus has, however, decided upon an immensely ancient origin for the Earth rather than a recent one. The intellectual basis for this idea emerged between the end of the eighteenth century and the middle of the nineteenth, and it was first postulated by the Scot, James Hutton (1726–97). In 1795, he published a classic treatise on geology, *The Theory of the Earth*, which began the inexorable backward projection of our planet's origins in time.

Hutton suggested that the processes that had made the layered sedimentary rocks, thousands upon thousands of feet thick, could still be observed. He thought that the deposition of sediments at the bottom of the oceans, and their subsequent compression by later layers of sediment, could have given rise to the existing rocks. Since he knew the slow rate at which contemporary deposition processes were taking place, Hutton deduced that the gigantic thickness of sedimentary rocks implied the passage of correspondingly immense amounts of time.

All at once, 6,000 years of change was pathetically too small a slice of time to have accounted for the construction of the Earth's rocks. Subsequent studies and increasingly sophisticated techniques have enabled direct aging of the rocks. Using established rates of decay of specific atoms in rocks, their time of construction can be calculated, and these calculations have provided an unambiguous framework of rock ages, which stretches back not 6,000 but more than 4 billion years.

This chronology, particularly of the sedimentary rocks, also provides a chronology of fossils. The remains of living things, trapped in the sediments as they were laid down, usually confirm that a

Interpreting a fossil

Reconstructing the appearance in life of a long-dead fossil skeleton is a skilled, multi-stage process. Although paleontologists know more or less where fossils are likely to occur, it is often mere chance that turns one up. Once discovered, the fossil must first be exposed and cleared of the sedimentary rock in which it is embedded. Next, the distortions caused by compression, bending and disruption of the fossil in the rock must be understood and reversed by the reconstructor. Then, with knowledge of the positions of muscles and other soft parts, gained from attachment scars on the bones and comparison with living creatures, the body can be "fleshed out".

In the case of the primitive bird *Archaeopteryx*—whose fossil remains were found at Sölnhofen, southern Germany, in fine-grained limestone of the late Jurassic 195–135 million years ago—even the feathers can be added, since the fossil shows their imprint.

rock and its fossils originated at the same time. The collection of creatures preserved in a specific layer of rock can, therefore, establish its position in the lineage of time with considerable precision.

The fossils in a particular rock stratum can act as its "brand mark". One of the most important scientists on exploratory oil-drilling rigs is the micropaleontologist—the expert on minute

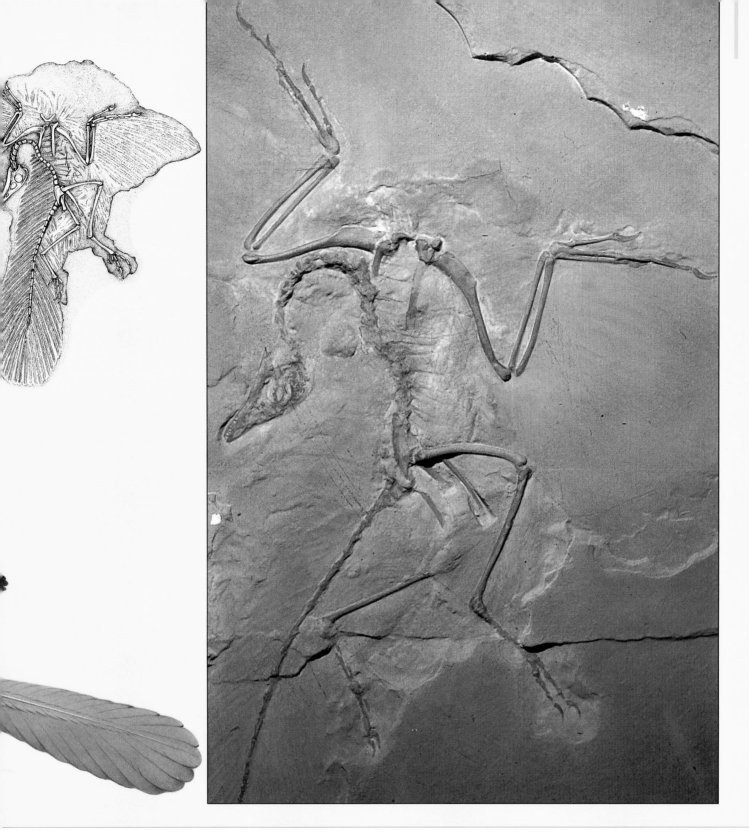

fossils of planktonic creatures. As cores of bored rock came up from thousands of feet below the ground or seabed, that fossil expert, using technology such as scanning electron microscopes, can date the rocks by their fossils with great exactitude. The type of fossil can also provide clues as to whether the rocks are likely to be near an oil-bearing seam.

The fossil record is a wonderful book of past life forms; but it is a book written in only a few "languages" and with many pages missing. The gaps in the book's coverage are caused by a variety of factors, and they ensure that there will always be some uncertainty and mystery concerning the picture fossils give of past life.

In vast areas of the world, much of the sedimentary rock from particular periods has been eroded away. The fossils that were in those rocks have been utterly destroyed, and we can never reconstruct them. Such destruction is a form of "fossil extinction"—the fossils disappear from the face of the Earth. Erosion is not the only destructive force of this type. Fossils are nearly always obliterated when sedimentary rocks are subjected to extreme heat and pressure and transform into metamorphic rocks.

The "languages" of fossils stem from the situations in which sedimentary

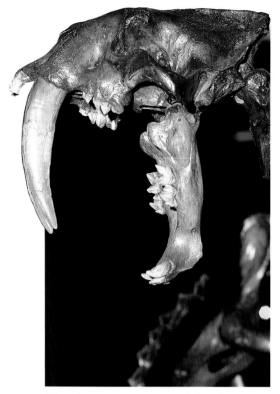

More than 2,000 specimens of the sabre-toothed cat, Smilodon californicus, *have been found at La Brea. A massive head and neck, and a jaw that opened to over 120°, enabled* Smilodon *to drive its huge curved teeth deep into its victim's flesh.*

deposition occurs. This takes place mainly on sea and lake beds, so there are innumerable fossils of underwater creatures of the past. It also happens in landscapes of windblown sand, so there are fossils from arid regions. But in a rain forest, for instance, deposition does not take place, since most dead organisms are rapidly eaten or decay. Fossils from such habitats are rare.

To have a real chance of securing a page in the book of fossils, an organism must have hard parts. The soft tissues of animals and plants rapidly break down, and it is usually only shells, cuticle, teeth, bones and other hard skeletons which remain to form fossils.

But the blanks in the fossil history of our planet are never total. There are always what one paleontologist has called "geological miracles". These are the wholly exceptional circumstances that from time to time have allowed soft tissues to be preserved with considerable fidelity to their original form. They are also the chance occurrences which mean that we have beautiful fossils of

animals which must have had only a miniscule chance of ever becoming part of a sedimentary process.

One such example of fossil perfection is the Burgess Shale in British Columbia, Canada. This extraordinarily old layered rock dates from the Cambrian age, and the wonderful animals we can see with diagrammatic clarity within its thin layers are more than 500 million years old. The extremely fine-grained shale has preserved both the hard and soft parts of many invertebrate animals.

Trilobites in this rock, for instance, are so well preserved that we can see their every appendage, the facets of their eyes and even, in some cases, what they had in their guts the day they died half a billion years ago. Some of the creatures in the shale cannot be assigned to any animal grouping known today. They appear to be early "experiments" in life forms that soon became extinct, with no persisting relatives.

Different types of chance fossilization have created other geological miracles. There are, for instance, the tar pits of Rancho La Brea in Los Angeles, formed some 25,000 years ago when crude oil seeped to the surface and, as its volatile components evaporated, left behind sticky pools of tar. Water lying in these innocent-looking pools must have drawn animals such as giant sloth and elephant to drink, and their struggles, in turn, probably attracted the carnivores: sabretoothed cats, dire wolves and vultures. All were trapped in the tar and died there, their bones immune to all normal processes of decay.

Then there are perfect insects, spiders and even tree frogs set in amber 70 million years old, which was originally resin from a tree. At the other end of the fossilization time scale, from the last ice age only thousands of years ago, are the frozen mammoths of Siberia, whose hide and hair is still intact.

A glass-bottomed boat or a mask and flippers are simple aids which can reveal the incredible living world of a coral reef to us. In the same way, fossils in the rocks of the Earth can open up, for all who want to see, the no less mysterious and fascinating living world of the past.

Some of the most beautiful fossils are those of insects which became trapped in the resins from trees, such as this 500-year-old Dipterus *fly. Even the fly's diaphanous wings and huge eyes have been preserved by the copal.*

The wonderfully well-preserved fossil of Thrissops formosus, *with scales and fins almost intact, dates from the end of the Late Cretaceous c65 million years ago. This 60-cm/2-ft streamlined predator was probably an ancestor of modern, freshwater, bony-tongued fish such as the North American goldeneye,* Hiodon alosoides.

Why did the dinosaurs die?

Extinction is a fact of life. No year goes by without some of the millions of species of organism on our planet becoming extinct by the death of its last member. But it is certainly the mysterious and mighty loss from the world's fauna of those prehistoric reptiles, the dinosaurs, that has most firmly grasped the human imagination.

The manner of the departure of the dinosaurs has prompted theory after theory. Some notions are carefully interleaved with all the known fossil facts. Others seem to be driving down the fast lane of pure speculation, with only the merest glance in the rearview mirror of evidence.

Extinction normally affects only those species least able to cope with the challenge of changing conditions or new competitors—and, in evolutionary terms, it has the advantage of making way for new, better adapted species. But a very different extinction event took place around 66 million years ago, at the time when the Cretaceous period, the Age of Reptiles, gave way to the Tertiary period, the Age of Mammals.

The extinction affected a wide range of organisms, both on land and in the sea, but it was most dramatic in the case of the dinosaurs. In all, 19 dinosaur families disappeared, as well as both families of leathery winged flying reptiles, the pterosaurs. The extinctions in the oceans were no less devastating. The ammonoids, which were like the modern nautilus and had existed for more than 300 million years, became extinct. So did the great marine lizards known as mosasaurs, many types of sea lilies, sea urchins, reef-building corals, mollusks and more than half the tiny planktonic marine organisms.

Nor, so the fossil record reveals, were the extinctions confined to animals. The diversity of plants is shown most clearly by their spores and pollen,

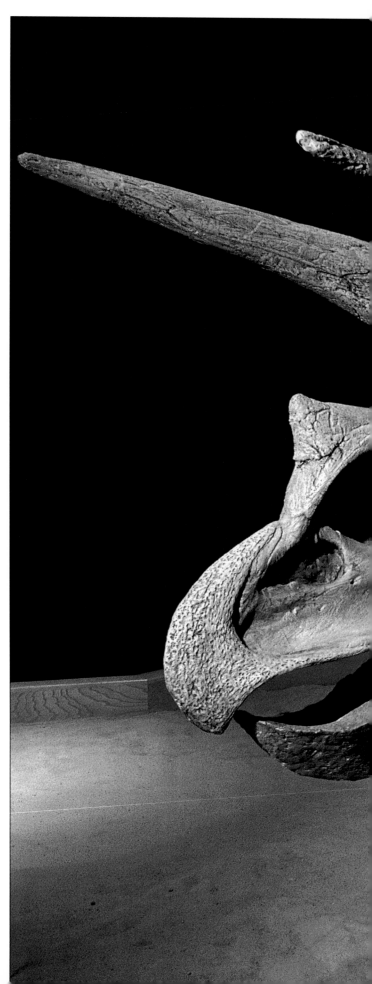

The ponderous, yet still majestic, fossil skeleton of Triceratops *makes us supremely aware of the catastrophic destruction of the dinosaurs.*

Theory upon theory has been advanced to try to explain the perennial puzzle of what wiped out these great beasts—and many of the Earth's other living creatures—after a reign of around 140 million years. Clues are gradually being turned up that clarify some aspects of the disaster, but the central mystery remains to baffle and intrigue.

which are better preserved than their fragile leaves and flowers. In North America, the rock strata laid down at the Cretaceous/Tertiary boundary (known as the K/T boundary) shows the sudden disappearance of 75 percent of the types of flowering-plant pollen, and an increase from 15 to 99 percent in the amount of fern spores.

So dramatic a pattern of extinctions could not have been simply the result of the normal processes of evolution and competitive replacement. Some more fundamental change must have occurred, and it was natural for early researchers to seek this in an abnormally rapid and violent alteration in climate. Indeed, there is evidence that such an upset did take place.

At the K/T boundary, plants bearing broad, simple leaves characteristic of warmer environments became rarer, while leaves with a jagged, or saw-toothed, outline characteristic of cooler environments became more common. Studies of marine sediments reveal that the amount of plant plankton also dropped sharply, another indication that temperatures became cooler. And if there were less plankton in the sea, that might well explain all the other extinctions of marine animals higher up the food chain.

Some scientists have pointed an accusing finger at the Moon as the cause of these extinctions. Because of their prevalence in rocks of that age, it is surmised that volcanic activity on the Moon sent out showers of microtectites at the end of the Cretaceous period. When these tiny glasslike spheres impinged on the Earth's atmosphere, they could have blocked out the sunlight sufficiently to produce massive cooling,

administering the *coup de grâce* to the dinosaurs.

The first real direct evidence for a possible cause came in 1981. In that year, Luis and Walter Alvarez, of the University of California at Berkeley, were studying sediments in central Italy laid down at the Cretaceous/Tertiary boundary. They found a layer of clay 13mm/$\frac{1}{2}$in thick which contained an unusually high concentration of iridium and osmium. These elements, normally rare in rocks originating on Earth, are abundant in meteorites.

Investigations in other parts of the world have revealed that rocks of the same period, in areas as far apart as Denmark, New Zealand and the Pacific Ocean floor, also contain high levels of iridium and osmium. Furthermore, the boundary clay in these terrestrial sediments shows 100–10,000 times the normal quantities of soot, and sometimes charred plant material, suggesting that there may have been devastating forest fires at that time. Finally, the sediments contained evidence of an impact, in the form of grains of quartz and other minerals showing sets of fine, intersecting lines, which are found in rocks that have been the subject of violent collision.

This volume and variety of evidence suggests that the K/T extinctions were the result of the Earth's having been struck by one or more great meteorites. The impact would have started forest fires in the tangle of fallen vegetation and would have hurled clouds of dust and

debris into the atmosphere. This, in turn, would have blacked out the sun and altered the climate for many years.

If this theory is true, where is the great crater that must have been formed when the meteorite fell to Earth? If there were only a single enormous meteorite, it would have had to be about 10km/6mls in diameter to produce the amount of iridium distributed around the world, and such a body would have made a crater about 100km/62mls across. The mineral grains showing the results of violent impact are most abundant in North America. So it is logical to seek there first for the crater caused by that fateful asteroid.

Meteor Crater in Arizona, initially an

Thescelosaurus

Many families became extinct at the K/T
boundary. Among the mosasaurs, the marine
lizard Platecarpus disappeared, as did the
dinosaurs Thescelosaurus, a herbivore that
lived in herds as modern deer do, and
Tyrannosaurus, the largest carnivore ever
known. The pterosaurs, flying reptiles such as
the pelicanlike Pteranodon, with a 7-m/23-ft
wingspan, also vanished.

Pteranodon

Platecarpus

Tyrannosaurus

Cretaceous		Tertiary % of families remaining
Dinosaurs		0%
Pterosaurs		0%
Terrapins		100%
Crocodilians		67%
Marsupial mammals		33%
Placental mammals		75%
Other mammals		75%
Lizards and snakes		107%
Mosasaur lizards		0%
Plesiosaurs		0%
Ammonites		0%
Sea urchins		71%
Bivalves		92%
Plant plankton (genera)		47%
Zooplankton (genera)		13%

The Deccan Traps

A crucial piece of evidence to prove the "giant meteorite" theory of the dinosaurs' demise is yet to be found: the location of the impact site. If this were in a part of the sea floor, it would have been consumed at the destructive margin of a crustal plate during the past 66 million years. But if the meteorite had impacted on land, the crater would be a localized feature, in which the effects of the fall—shock-altered quartz, soot from vast fires and an increase in heavy metals common in meteorites—would be found.

A strong contender for the site is the Deccan Traps in India, an area of plateaulike steps of solidified lavas dating precisely from the K/T boundary. But where is the crater, which we might expect to be 100–200km/62–124mls wide? Perhaps it has been obliterated by the massive volcanic activity produced by so great an impact.

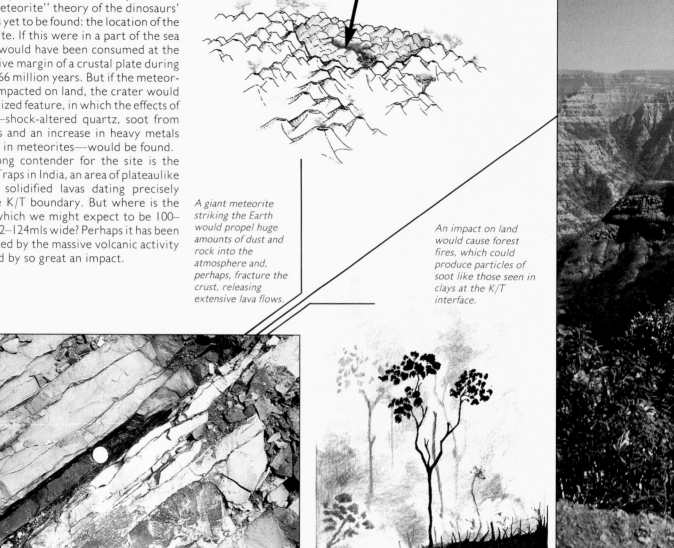

A giant meteorite striking the Earth would propel huge amounts of dust and rock into the atmosphere and, perhaps, fracture the crust, releasing extensive lava flows.

An impact on land would cause forest fires, which could produce particles of soot like those seen in clays at the K/T interface.

Clay layers at the K/T boundary contain soot and heavy metals.

obvious candidate, is, however, only a few thousand years old. But another possible crater of the correct age has been identified near the town of Manson in Iowa. Since it is covered with 30–90m/98–295ft of sediments laid down later by ice-age glaciers, it is difficult to study directly, but with a diameter of only 40km/25mls, it seems too small if the great extinctions were caused by a single meteorite impact.

Although a meteorite may well have produced the iridium and shocked minerals of the K/T boundary, some scientists believe that the boundary clay and osmium may have come instead from

volcanic eruptions. This theory is supported by the discovery of volcanic ash at the boundary level at several sites, as well as other rare minerals which are usually found at high concentrations only in volcanic rocks. More importantly, one of the greatest volcanic areas in the world, the Deccan Traps in India, appears to have been formed by an eruption at the K/T boundary.

Perhaps the reason for the eruptions in India was that the meteor struck there, penetrating 20–40km/12–25mls into the Earth and exposing the molten rocks of the Earth's interior. These may then have welled up to flood the surface

with great outpourings of lava. This theory would explain most of the changes that appear to have taken place about 66 million years ago, and it seems an advance on older, vaguer theories.

Most fundamental of the questions still left unanswered is just how suddenly the changes occurred. It is difficult, even at a single location, to estimate precisely the length of time represented by the deposition of sediments around the K/T boundary. And it is even more difficult to know how that time relates to the period during which the extinctions took place on land and in the sea. Did these take 100 years, or 10,000? And did

The volcanic Deccan Plateau covers an area of 518,000sq km/200,000sq mls.

the extinctions take place at the same time in North and South America and Australia? Because it is not possible to date the sediments absolutely accurately, we cannot be certain that the last rocks containing dinosaur fossils are the same age in all the continents, and it may be that the suddenness of the dinosaurs' demise has been exaggerated.

While subscribing to the giant meteorite theory, some scientists have suggested that it caused not a "nuclear winter" but, rather, rapid global warming. If the meteorite had impacted on limestone rocks, it could have released huge amounts of carbon dioxide into the atmosphere, and this would have caused an immediate and violent greenhouse warming of the planet. Perhaps it was this that put paid to the giant reptiles.

Yet another theory proposes that it was problems with the ratio of the sexes among dinosaurs that was the root cause of their extinction. We know that in some modern reptiles, such as the crocodile, the sex of offspring is partly determined by the external temperature while the eggs are incubating. If there were rapidly cooling conditions 66 million years ago, they might have distorted the dinosaurs' sex-determination mechanisms to produce a female-biased population. Their reproductive success would have dropped, since many would not have found mates.

The theories, speculations and ideas as to what caused the dinosaurs to vanish seem never to end. But we can probably be certain of only one thing. Since the event took place the equivalent of a million human lifetimes ago, it is most unlikely that we shall ever understand the true reason for its occurrence. The mysterious loss of these extraordinary reptiles, which held sway over the Earth for so long, will continue to tease us as a reminder of the inevitability of our own ultimate extinction.

In search of our beginnings

The Earth today is ruled by *Homo sapiens sapiens*, "Modern Man", who is quite unlike any of the earlier species to dominate our planet. The dinosaurs ruled the world for millions of years, but they depended for survival on the nature of their environment. Humans have altered the landscape, adapted in such a way that climate no longer restricts their range, and have even changed the chemistry of the Earth's atmosphere to an extent that may cost their very existence. It is, then, surprising that the story of such a species as our own should be so hard to decipher.

The search for our ancestry takes us to Africa, where fossil evidence has been found that extends our lineage back some three million years. It is in areas that we now call Ethiopia, Kenya, Tanzania and South Africa that the remains of hominids of the genera *Homo* and *Australopithecus* have been uncovered. It seems certain that the evolving hominids eventually spread out from this geographical "cradle of mankind".

The human lineage can, however, be traced even farther back. As primates we share a common ancestry with the lorises, lemurs, tarsiers, monkeys and apes, and the origins of the last of these stretches back at least 38–24 million years. It is still not certain when the hominids—which include the ancestors of "Modern Man"—split away from the apes.

An apelike creature, *Ramapithecus*, is the most likely key to this mystery. Named for the god Rama because it was first found in the Siwalik Hills of north India in 1931, this fossil, and others like it, are of Miocene age, between 24 and 5 million years before the present. Their small canine and incisor teeth and squat lower jaw, together with the fact that the bony arch of the cheek is farther forward, give it a close resemblance to later hominids. As with the ancestors of

It seems fitting that Africa, over whose landscapes hangs such an air of timelessness, should have been the first home of nature's most amazing creation.

At sites all along the Great Rift Valley in East Africa, one of Earth's most exciting mysteries is being teased out of ancient rocks. In the Olduvai Gorge, for instance, the river has cut down through the red and black lavas to expose sediments in which have been found the fossilized remains of Zinjanthropus, a hominid one and three-quarter million years old. Yet on the surrounding plains of the Serengeti roam animals little different from those such early beings must have known.

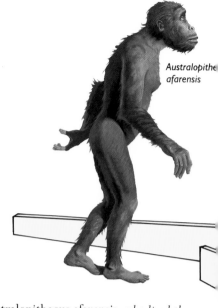

Australopithecus afarensis

the apes, *Ramapithecus* roamed a large area: specimens have been found as far apart as Macedonia, southern China and East Africa.

Two main factors probably brought about the restriction of the later hominids to Africa. The Pliocene epoch saw a massive shift in the climate, culminating in the Pleistocene ice ages, which began about two million years ago. This caused the subtropical and tropical forests to dwindle in extent, and with them the type of habitat that supported the ramapithecines.

The expansion of more open habitats may have created the ecological niches into which fast-evolving species, including the ancestors of humankind, could expand. But the range of these hominids, from roughly 11 degrees North to 27 degrees South, indicates a preference on their part for a warm climate.

The other reason for the geographical concentration of the early hominids may have been entirely fortuitous. An ancient continent such as Africa is planed flat by the forces of erosion over much of its area. The number of places where sediments bearing the fossilized remains of animals can accumulate is not extensive. As it happens, many of the hominid fossil sites are associated with hollows such as the Great Rift Valley, into which sediments have been washed by rivers.

Other fossil sites, specifically those of South Africa, are caverns, into which it is thought the hominids may have been dragged as the prey of carnivores such as the big cats. As in the Rift Valley, sediment buried the skeletons and so helped to preserve them until erosion laid them bare for paleoanthropologists—those who study fossil humans—to discover.

The distribution of the hominids was, almost without doubt, wider than just the Rift Valley and a handful of caves. So why are the remains of our forebears so scarce? The answer can be gleaned from the fate of animals that die on the present-day plains of Africa. Much of a carcass is dismembered by scavengers, so parts of the body, such as the limbs, are scattered. And because the bones are

not buried by sediments, within a few years they disintegrate and so do not become fossilized.

Lack of evidence, and the widespread rejection of the idea that apes and humans share a common ancestry, make the history of the early hominids difficult to unravel. The elaborate hoax in 1908 of the ''Piltdown Man'', with its modern skull and ape jaw, also contributed to make the scientific community sceptical. As a result, extreme caution greeted the discovery in 1924, by the South African fossil expert Raymond Dart, of what was thought to be a child's skull near Taung in the Transvaal.

But this find, together with those in caves at Sterkfontein and Kromdraai in the 1930s, and at Makapansgat and Swartkrans in the 1940s, was an important breakthrough. The specimens were hominids, although not of the genus *Homo*. The smaller ones were named *Australopithecus africanus*, ''Southern Ape of Africa'', the larger were initially named *Paranthropus robustus*, ''Beside Man'', and were thought to have evolved from *A. africanus*. Some are now considered to be merely large and small species of the same genus, and the name *Australopithecus* is used for both. The impossibility of dating the cave sediments in which these hominids were buried means that their evolutionary pattern will always be speculative.

This was less of a problem at Olduvai Gorge in Tanzania. Sediments had been laid down in and at the margin of a lake, which had swollen and shrunk with changes of climate. With the lake now dried up, and the sediments incised by a downcutting river, layers of deposits have been exposed in the valley sides. It was here in 1959 that the anthropologist Louis Leakey (1903–72) and his wife Mary found ''Zinj''—*Zinjanthropus*—later reclassified with the robust australopithecines of South Africa. ''Zinj'' was dated at about one and three-quarter million years ago, as were other australopithecines at Peninj, near Lake Natron in Tanzania, and at Chesowanja, near Lake Baringo in Kenya.

Then, during the 1970s, the American scientist Don Johanson and his team,

Australopithecus afarensis, who lived three to four million years ago, is the earliest known hominid. From this stock, over the next million years or so, two types evolved, including the heavy-jawed A. robustus *and the more lightly built* A. africanus, *ancestor of the genus* Homo. *From about two million years ago,* H. erectus *and the australopithecines coexisted, until the latter died out. Our own species,* H. sapiens, *dates from less than 250,000 years ago.*

working at Hadar in the Afar Triangle of Ethiopia, found several hominids with characteristics distinct enough to suggest a separate species, *Australopithecus afarensis*. In the case of ''Lucy'' (named for the Lennon/McCartney song *Lucy in the Sky with Diamonds*), where just under half the skeleton was recovered, the pelvis and leg bones indicated a bipedal stance. The sediments in which Lucy rested were dated at just over three million years; this was the first evidence that hominids moved on two legs as far back as the Upper Pliocene.

Other exciting finds were to follow quickly. Mary Leakey, searching at Laetoli near Olduvai in Tanzania, chanced upon what paleontologists call a ''trace fossil'', usually an impression left in soft sediment. In this instance, recently deposited volcanic ash had been wetted by a rainstorm and had the consistency of freshly poured cement when three hominids crossed it, leaving a trail of footprints. These confirmed as

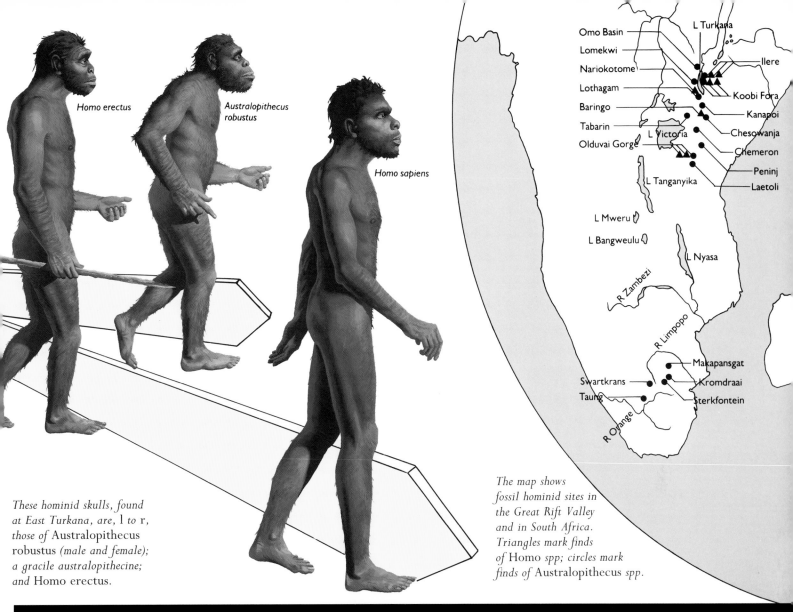

Homo erectus

Australopithecus robustus

Homo sapiens

These hominid skulls, found at East Turkana, are, l to r, those of Australopithecus robustus (male and female); a gracile australopithecine; and Homo erectus.

The map shows fossil hominid sites in the Great Rift Valley and in South Africa. Triangles mark finds of Homo spp; circles mark finds of Australopithecus spp.

Omo Basin
Lomekwi
Nariokotome
Lothagam
Baringo
Tabarin
Olduvai Gorge

L Turkana
Ilere
Koobi Fora
Kanapoi
Chesowanja
Chemeron
Peninj
Laetoli

L Victoria
L Tanganyika
L Mweru
L Bangweulu
L Nyasa

R Zambezi
R Limpopo
R Orange

Makapansgat
Swartkrans
Kromdraai
Taung
Sterkfontein

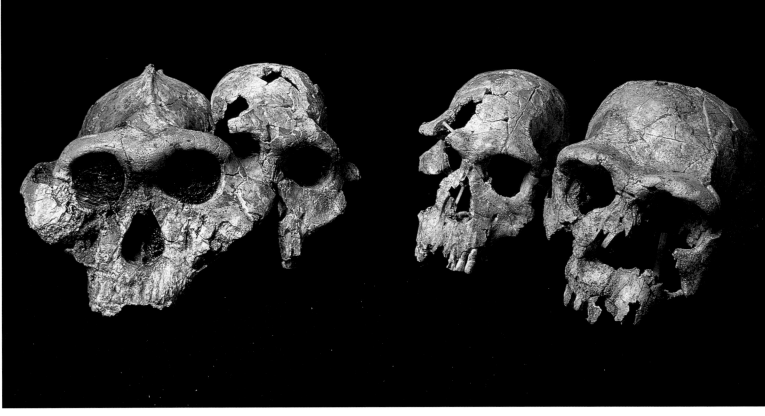

positively as any anatomical evidence could have done that hominids had an upright, bipedal gait three and a half million years ago.

These footprints were, without doubt, made by australopithecines, specimens of which have been found in sediments of the same age. But what of *Homo*?

In 1961, at Olduvai, fragments were pieced together of another hominid, whose cranial capacity of 800cu cm/49cu in was much bigger than that of all the other australopithecines. Louis Leakey, for some time puzzled by the stone artefacts that had been found at the same sediment level as "Zinj", now thought he had evidence of the superior intelligence responsible for these tools. He named this new creature, dated at just under two million years, *Homo habilis*, "Handy Man".

Attention then switched north to another part of the Great Rift Valley. Richard Leakey, son of Louis and Mary, returning to Nairobi from an international dig in the Omo Valley, and flying down the east side of Lake Turkana, decided that an area marked cursorily on the map as "lava" was nothing of the sort. In 1968, he led a preliminary expedition to the area, which established the Koobi Fora basin as a prime hominid fossil site.

In 1972, the more or less complete cranium of "Skull 1470", another specimen of *Homo habilis*, was discovered here. Originally believed to be older than those from Olduvai, it is now regarded as comparable in age.

Koobi Fora has been a prolific site, of considerable importance in tracing the evolutionary lineage of humans. In particular, evidence of the coexistence there of robust australopithecenes of a type similar to "Zinj" and specimens of *Homo habilis*, such as "Skull 1470", puts paid to earlier theories that *Homo* evolved from *Australopithecus*. It has also been the location of another important discovery, the remains of *Homo erectus*, "Upright Man".

Examples of this link fossil are known from other sites as far apart as Europe and Indonesia. What makes the Koobi Fora specimen important is its age, since it places *Homo erectus* firmly in East Africa at one and a half million years ago, far earlier than those elsewhere.

Homo erectus had a more sophisticated

All primates have opposing thumbs, enabling them to grasp objects. Monkeys and apes, such as orangutans, also have opposable big toes, while humans, walking upright on two feet, are distinguished by having little flexibility in their big toes.

tool kit than *Homo habilis*. For example, the site at Olorgesailie contained some stone handaxes, and round stone tools which may have been used to bring down game in much the same way that South American gauchos use the *bolas* to floor their cattle.

So why did *Homo erectus* succeed *H. habilis*, and why did the australopithecines eventually become extinct? The answer may lie in environmental change. The geological record shows that the climate of East Africa became significantly drier about two million years ago, as the ice began to advance in northern latitudes. Vegetation became tough and leathery, and open grassland expanded at the expense of trees.

The creation of different ecological niches meant success for the species that could adapt fastest, just as the disappearance of much of the old habitats spelt doom for species that were slow to adapt. Microscopic examination of the scratch patterns on the teeth of *Homo habilis* and *Australopithecus robustus* shows that they are well polished, suggesting fruit-eating and excluding grass-eating and bone-crunching. In contrast, the teeth of *Homo erectus* were much scratched and indented, indicating an omnivorous diet which included roots carrying soil. Their improved collection of tools and possible weapons suggests active, organized hunting and increased meat-eating. So it would appear that *H. erectus* succeeded where *H. habilis* and *Australopithecus* failed by being able to exploit the harsher conditions imposed by the drier climate.

This is not the end of the story of human evolution, but it seems to be the end of that part which is specifically African. With a larger and still developing brain, with such technical facilities as vastly improved stone tools and the use of fire, and with a growing ability to exploit harsher, seasonal environments, *Homo erectus* now stepped out of the cradle and migrated both north toward Europe and east toward China and the East Indies. There is much subsequent history still to unravel, and there is no reason to suspect that *Homo sapiens sapiens* is the final form of our genus.

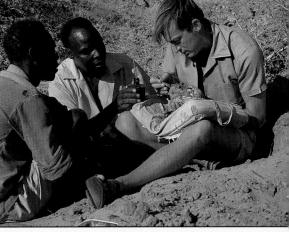

Mary Leakey's find of fossilized footprints at Laetoli in 1978 proved that hominids, probably A. afarensis, were bipedal some three and a half million years ago. The prints, which appear to be those of two adults and a child, show a well-developed arch to the foot and no divergence of the big toe. Richard Leakey, right, with his skilled assistants, has greatly increased our knowledge of our human lineage with his discoveries, notably at Koobi Fora, where Homo erectus was found.

New life
on new land

All life on Earth is interconnected, in a sense. And within communities of living things life forms seem to operate so closely together that they are almost like incredible ''super-organisms''.

A tropical rain forest is one such ''super-organism''. Here, the trees are the primary producers, or sustainers, which trap sunlight with the green chlorophyll in their leaves and use it to make living substance or biomass. This they do by the process of photosynthesis, capturing carbon dioxide from the air and combining it with water to make new living material.

In the forest, myriad herbivorous animals—from leaf-eating monkeys to caterpillars—depend on these plants for food. Carnivores, from spiders to monkey eagles, are inextricably connected to the herbivores by predator-prey linkages in the food chain. Decomposers, including bacteria and fungi, break down the dead bodies of all these life forms to recycle their components.

Other cross-links appear at every juncture. The trees are a living scaffolding for epiphytic ferns, lianas and orchids, and the nests of birds; the insects, birds and bats are essential pollinators for so many of the trees; and monkeys and parrots eat their fruits and disperse their seeds. Link upon link, connection within connection, a concatenation of life exists.

There is, however, a weird and mysterious paradox nested inside these layers of biological interdependence, a central and vital exception to the ecological rule. New forms of life seem only to form when barriers and isolation come into the scheme of things. Incongruously, communities can operate only because of their interconnections; yet they can survive long-term only by generating new, better-adapted life forms in response to changing environmental conditions.

Oceanic islands, formed by underwater volcanic eruption, emerge above the sea devoid of any form of life—they are truly desert islands.

With amazing rapidity, however, living things take hold. Seeds and plant spores arrive from afar with the sea or the wind, or are ferried in by birds, which also remain to colonize the new land. Such isolation provides ideal conditions for the mysterious forces of evolution to act upon these living forms, giving rise to species immaculately tailored to the environment.

Charles Darwin was a naturalist and student of theology, and it was only after many years of deep thought and research that, in 1858, he published his seminal work on the origin of species that so shocked the Victorian world. This was followed in 1871, when he was 68 years old, by the even more controversial, The Descent of Man. Darwin was widely reviled in his time, but his work is now regarded as having opened the door to present-day understanding of the natural world.

The teeming wildlife of the Galapagos provided sailors with fresh food—easily come by, for the animals had little fear of humans. This picture from the Illustrated London News of July 1850 shows a sailor about to overturn a tortoise with a boathook; a common way of securing a meal.

The iguanas Darwin saw on the South American mainland climbed trees and ate leaves. On the islands, where vegetation was sparse, although the land iguanas below, climbed cactus trees to feed on young, juicy growth, one species had become marine. These iguanas were able to swim and lived on seaweed.

Darwin and the Galapagos

The Galapagos—13 large and many small islands—lie on the Equator, some 1,054km/ 650mls off the coast of Ecuador. They were named in 1535 by the Spanish navigator Tomás de Berlanga for the giant tortoises he found there. It was these animals, together with the strange marine iguanas and many species of finch, which vary from island to island, that prompted Darwin's speculations on species specialization. Tortoises living on dry islands have long necks and a high peak in their shells, enabling them to stretch up to browse on cacti and bushes growing there. On islands where there is ground vegetation, adaptations such as these are absent.

It seems that this process of change happens most frequently when populations are in some way isolated from one another. So the natural history of islands plays a central part in enabling us to grasp the core of the conundrum, since they are nature's own isolation laboratories. For any terrestrial organism, particularly one that cannot swim or fly well, an archipelago of islands is a habitat divided by practically impenetrable barriers—the stretches of open water between them.

The population of a species on one such island may be only tens of miles from the population on an adjacent island, but the intervening ocean ensures that there can be no cross-breeding and mixing of hereditary material. With slightly different environmental conditions on the two islands, and slightly differing patterns of competition, the evolutionary paths of the two populations will steadily diverge as each becomes better adapted for its own local circumstances.

Any pattern of geographical barriers can give rise to virtual islands. If, notionally, water and land "change places" and an archipelago becomes a cluster of lakes embedded in a continent, their aquatic life will be subject to exactly the same relative patterns of isolation, lake to lake, as land animals experience on islands. Limestone outcrops in an acidic landscape form "islands" for lime-loving animals and plants, while mountain peaks in the tropics are "islands" for alpine creatures in an otherwise equatorial climate.

The very idea of new forms of life being possible—the concept of evolution—owes much to the ecology of islands. In 1859 a book was published whose first edition sold out on the first

day of publication; by 1860 it had run through three further editions. This was no popular novel, nor an improving tract. It did not have an obviously arresting title: *On the Origin of Species by means of Natural Selection, or the Preservation of Favoured Races in the Struggle for Life* does not, on the face of it, sound like an instant best-seller. But the author of the book was Charles Darwin (1809–82) and the problem it confronted was one of the major mysteries of the life sciences—where do new types of living things come from?

Darwin's seminal thoughts on the

matter had crystallized only slowly, for the seeds were planted during his voyage on the surveying and exploration vessel HMS *Beagle*, which circumnavigated the world for the British Admiralty between 27 December 1831 and 2 October 1836. And their germination was brought about by the ecology of the Galapagos Islands in the Pacific Ocean.

The *Beagle* was moored for just over a month in the Galapagos Islands, from 16 September to 20 October 1835, but it is not too much of an exaggeration to say that those 35 days generated a major upheaval in human thought. The study of

the animals there made Darwin realize that isolation and selection for new adaptations could generate new forms of living things. Darwin himself was in no doubt eventually about the crucial nature of his Galapagos experiences. He was later to say of the islands: "Here, both in space and time, we seem to be brought somewhat near to the great fact—that mystery of mysteries—the first appearance of new beings on this earth."

What Darwin found on each island was an extraordinary community of animals, almost all of them unique: they existed nowhere but on the Galapagos. Their nearest and most similar relations were all to be found on the South American mainland hundreds of miles to the east. What is more, many of the animals—particularly the finches, later to be called Darwin's finches, and the giant tortoises—had clearly different forms on each of the tiny volcanic islands of the Galapagos group, the largest of which is only 129km/80mls across.

This profligacy of novel life forms in such a small geographical compass amazed Darwin. He noted also the extreme adaptations of so many of the creatures—a cormorant that had become completely flightless; an iguana lizard that dived into the sea to collect seaweed for its food; a finch that used a cactus spine as a probing "tool" in its beak to prise grubs out of holes.

After many years of pondering the facts, Darwin came to a startling and terrifying conclusion. The only explanation for the Galapagos phenomenon was that rare colonist animals had flown or drifted from the mainland to the islands over previous millennia. Once there, with little local competition, the colonists had become separately adapted to the special conditions prevailing on each of the different islands. They had changed to fit their environments. In a word, they had evolved.

Although the fact of these changes became evident to Darwin while he was on the Galapagos, it was only later that he was able to suggest a mechanism for them. The explanation he proposed was the "survival of the fittest". This means that within a population of animals or plants, each with a slightly different genetic make-up, those organisms best able to survive in the environment and adapt to the circumstances in which they find themselves will flourish. These survivors will live long enough to reproduce and pass on their "fit" genetic blueprint to the next generation.

Modern genetic research upholds Darwin's basic concepts and, subsequent to his astounding proposals, scientists have realized that clusters of isolated, mid-oceanic islands always show a similar pattern to that of the Galapagos,

although the particular organisms that are found will vary from group to group. The Hawaiian Islands, for instance, are the most isolated in the world—more than 3,000km/1,865mls from North America and 5,000km/3,100mls away from Asia—mere pinpoints of volcanic rock in the vastness of the Pacific Ocean.

The distances involved are so great, the vicissitudes of colonization journeys so enormous, that some animal groups have never made it naturally to the islands. There are no freshwater fish, amphibians, reptiles, land mammals or coniferous trees; none of them managed to make the trip. But the relatively tiny sub-set of groups that did make landfall had unparalleled colonization opportunities, with few competitors. They have exploited their evolutionary good fortune, expanding into a wide range of different, specialized types.

A perfect example of the way this adaptive radiation operates is found on the Hawaiian islands. A collection of finchlike birds reached the islands,

The emperor bird of paradise and all the other bird of paradise species inhabit the rain forests of New Guinea, the Moluccas and northern Australia. The six species of gibbon, including the lar gibbon, right, are purely Asian.

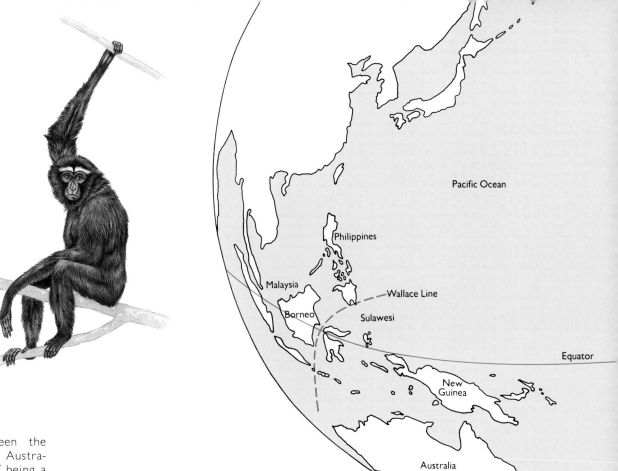

Pacific Ocean

Philippines

Malaysia

Wallace Line

Borneo

Sulawesi

New Guinea

Equator

Australia

Wallace's Line

The multiple islands between the Southeast Asian mainland and Australasia give every impression of being a consolidated archipelago. They are, however, riven by an invisible divide, known as Wallace's Line for its discoverer Alfred Russel Wallace (1823–1913), whose concept of evolution paralleled that of Darwin. The animal life of the different islands provides the clues by which the line can be located, for particular types are found only on the Australasian or Asian side of the line.

adapted to life in the many niches they had to offer, and evolved into the exclusively Hawaiian family of the honeycreepers. Now there are more than 40 distinct species, with habits specialized for a variety of lifestyles, feeding on insects, insect larvae, fruits, seeds and nectar. Some have even developed quite unfinchlike beaks: long curved probes used to extract nectar from long tubular flowers.

Among the plant seeds to arrive in the Hawaiian island chain were those of

The weird, scaly pangolin is found on the Asian side of Wallace's Line, as far west as Africa, while the common cassowary lives only in the forests of Australasia. Due to the isolating effects of island habitats, many races of this bird have evolved.

simple lobelias. In the absence of already established trees, some lobelias were able to evolve to fill even this niche, and several species now grow more than 9m/30ft high. In all, there are more than 100 species on the islands, none of them known anywhere else in the world.

The Hawaiian Islands have the highest percentage of endemic creatures (those found in only one place) of any region in the world. More than 90 percent of the flowering plants, 99 percent of the birds and effectively all the insects are endemic to the islands. Nothing could illustrate more clearly the pivotal role of isolation in the production of new forms of life.

Sometimes we get the opportunity to witness the early stages of the colonization of an island. This chance occurs either when a new volcanic island emerges from the sea (as with Surtsey off the coast of Iceland in 1963) or when a pre-existing island is sterilized by violent volcanic activity (as when Krakatoa in Indonesia erupted in 1883). In both sets of circumstances it is remarkable how quickly life colonizes or recolonizes virgin ground.

In this intial populating of an island, successful colonization is not merely a matter of reaching it. Populations of animals and plants must be sufficiently large to enable them to breed, and communities must have the appropriate ecological components to form a food-web. A predator, for instance, reaching an island will starve if there are no prey animals to eat.

Apart from simple plant forms, such as algae, mosses and lichens, which are spread by spores, the first green colonizers of islands are usually weed species with rapidly germinating seeds, and small omnivorous animals. Because they can fly or air-drift to the new land, there are always many types of insects and birds. Some specialist forms are even pre-adapted for island-hopping and are regularly among the first arrivals. Tropical examples of successful island colonizers include the coconut palm, whose huge fruits can float for long periods in the sea, surviving the salt, and can then germinate on beach sand.

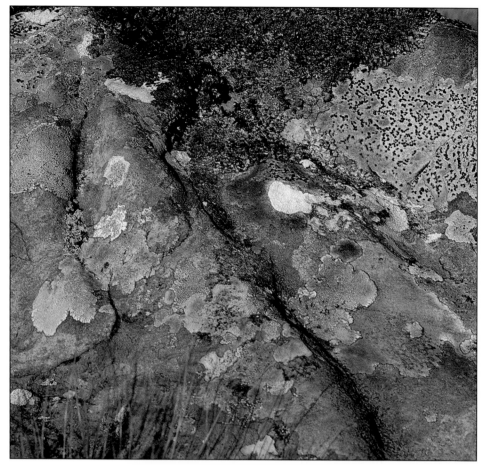

The first living things to colonize a new island are bacteria and single-celled green algae. They are followed by seaweeds and lichens, above, symbiotic organisms, consisting of algal and fungal partners, that appear to survive by the photosynthetic activity of the algae. Later, the spores of mosses, such as Bryum capillare, right, with its spore capsules almost ripe, reach the island.

Studies suggest that as the community of organisms increases on a new island, a form of ecological regulation sets in. There seems to be a "carrying capacity" of species dependent largely on the island's size—the bigger it is, the more species there will be. The species mix does not, however, remain static, and some forms vanish from the island to be replaced by new ones. The more species that arrive, the higher the risk of extinction because the competition is greater. In experiments on tiny islands it has been possible to show that if all the creatures are removed from an island and it is left to repopulate naturally, the

*Grass and rush seeds
are blown in by the
wind or carried on
drifting vegetation to
the island's shores.
And in the tropics,
coconuts, the only
seed impervious to
long immersion in salt
water, may arrive and
take root. As soon as
conditions are
favourable, insects
that arrive by chance
will survive and breed.*

final number of species will be much the same as the number present initially.

It is as though an island were a microcosm of a continent. As though the processes that are diffuse and confused in the massive complexities of continental life are condensed to diagrammatic clarity in the simpler life of islands. The

Galapagos presented these truths to Darwin and are proffering them to us still. By studying islands, particularly remote ones, we can begin to unravel the mysteries of evolution, the production of new living species, extremes of specialization, and the rules for life that underlie the ecology of communities.

*Seabirds, such as the
strong-winged, great
black-backed gull,
Larus marinus, are
among the first to
colonize any stack of
land in the ocean.
Birds bring with them
seeds and insects.*

The strange, still sea

In the western North Atlantic lies a strange, still sea, unbounded by land—the Sargasso. This oval mass of water, with an area roughly two-thirds as great as that of the continental United States of America, revolves more or less around the island of Bermuda. Those glorious beaches are the only land on which the waters of the Sargasso Sea lap.

Swirling ocean currents corral the waters of the Sargasso Sea: the Gulf Stream on its northern flank and currents streaming westward along the Tropic of Cancer on its southern edge. This impounding of the Sea results in a slowly revolving system of relatively warm water that rotates clockwise above the much colder waters of the Atlantic depths. Since it is less dense than the cold waters beneath it, the warm water forms a layer on the surface.

This density stratification of the waters, caused by temperature differences, has a profound ecological effect. Because plant plankton thrive in the well-lit surface waters, they use up salts such as phosphates and nitrates, and because of the density differential there is little mixing with the cold, mineral-rich waters below to enable their replenishment. So the upper regions of the Sargasso Sea are almost empty of rich marine life such as large fishes, and would be devoid of biological interest if it were not for the sargassum weed, which teems with living creatures.

It was far-ranging Portuguese sailors of old who, apparently, gave the weed and the Sea their names. Tangled patches of seaweed, buoyed up by gas-filled bladders, float everywhere, and in some places great rafts of weed stretch from horizon to horizon. Sometimes the Portuguese ships must have been slowed down, even enmeshed, by it, giving the crews plenty of time to study the weed. Coming from a country rich in vineyards, and probably missing the

In the mysterious, slow-moving waters of the Sargasso Sea, little fish dart among the fronds of green and gold sargassum weed as if they were flitting through the canopy of a rain forest. And as the trees of the rain forest nurture the jungle creatures, so the weed provides a totally self-sufficient habitat for the many life forms, some of them quite bizarre, that have taken up residence there.

wines of home, they likened the clusters of rounded gas bladders to bunches of grapes of a variety known as *salgazo*. And so the Sargasso Sea was named.

The weed derived originally from forms which grow attached to hard substrates near the shore. But now the sargassum weed is entirely pelagic—it floats in the upper layers of the open ocean only—and is able to remain on location and perpetuate itself there because of two factors. The currents slowly circulating around the Sargasso Sea keep most of the weed within its bounds, and the weed propagates by a form of reproduction based on fragmentation. It can form new plants in the same way that some land-based weeds, such as bindweed, do. Every tiny piece that breaks off a parent plant has the potential for regrowth.

The wide meadows of weed may appear pernicious but, in fact, they provide the biological framework to support a self-contained ecosystem dependent on the primary productivity of these plants. A complex mixture of encrusting organisms is attached to the creviced surfaces of the weed. There are small algae, hydroid coelenterates such as floppy coral, tube worms that sift the water for minute food creatures, colonies of "moss animals" or bryozoans, and many other forms.

Scrabbling over the weed fronds are crabs, prawns and shrimps, and other crustaceans; and tiny snails glide everywhere, rasping surfaces clean of microscopic algal growths with their rough tongues. The unexpected nature of this mid-oceanic menagerie fooled Christopher Columbus on his first transatlantic voyage. Noticing the sargassum crabs when still more than 1,610km/1,000mls out from America, he wrongly assumed that he was close to shore.

The fish that exist in this weed-world are particularly intriguing. Ecologically they can be divided into residents and migrants. Resident species live, breed and feed within the weed rafts of the Sargasso Sea and are often exquisitely specialized and camouflaged for life in the dappled olive-green seascape. The migrant species, the best-known of which are eels of the genus *Anguilla*, merely breed in the Sea, spending most of their lives far from the Sargasso, at the other terminus of countless incredible individual journeys.

The resident fish are of all types. Filefish, small relatives of the pufferfish, browse on the weed itself. Superbly camouflaged pipefish, with bodies like knitting needles, disappear among its stems, where they forage industriously for invertebrate prey.

But perhaps the most remarkable of these specialists is the sargassum fish, an angler fish whose scientific name, *Histrio histrio*, means "the actor". A consummate illusionist, its role is that of a weed frond. The fish's squat body, some 20cm/8in long, is blotched black and yellow-green to match the patterns and tints of the weed.

To cap this astonishing visual mimicry, the sargassum fish locks itself

The sargassum crab, Portunus sayi, *has adopted much the same camouflage as the sargassum fish. Such perfect adaptation as these creatures show suggests that the weed meadows of the Sargasso must have existed for a very long time.*

The slowly swirling currents of the North Atlantic both contain the waters of the Sargasso Sea and help the eel larvae on their journey to the coasts of North America and Europe. The ranges of these two populations are shown on the map on page 216.

Gulf Stream

Bermuda

Sargasso Sea

to the weed with amazingly modified pectoral fins so that it drifts, waves and floats exactly as the weed does. These fins have become short flexible "arms" on the ends of which are ten-fingered "hands"; the fin rays that make up the "fingers" are able to grip the weed firmly. When stalking prey, the fish does not swim but stealthily clambers through the weed, like some sloth

Tufts of skin, like fronds of weed, break up the outline of the sargassum fish. Its blotchy skin also has small white patches on it, which mimic the tube worms and moss animal colonies that thickly encrust the true weed.

An amazing life story

When the tiny, transparent leptocephalus (narrow-headed) eel larvae hatch, they drift for great distances across the ocean. As they approach the coast of North America after a year, or of Europe after at least three years, they transform into more eel-like elvers. Thousands of these little fish then swim upriver, where they continue to live and grow. When some 14 or 15 years old, the large eels lose their grey-green and yellowish coloration and become "silver eels", before making their long way back to the Sargasso Sea where they mate and die.

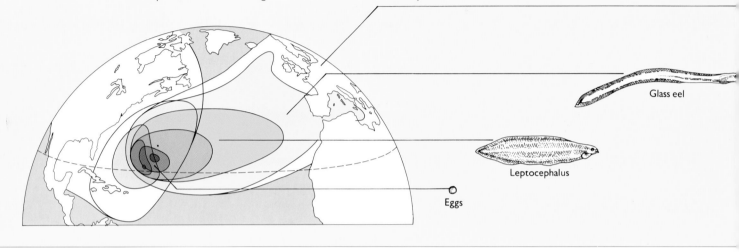

Glass eel

Leptocephalus

Eggs

moving relentlessly through the jungle canopy of Amazonia.

Prey, such as filefish and some of the larger invertebrates, are tempted close to the huge gape of the sargassum fish by a flexible lure—a frilled ray of the dorsal fin—which can be twitched enticingly in front of the fish's mouth. Prey animals, attracted to what appears to be a morsel of food, are rapidly engulfed, as the hunter violently opens its mouth and sucks them in.

But it is the mysterious life story of those long-distance migrants of the Sargasso Sea, the eels, that is the most intriguing. The fact that adult eels are common in freshwater in Europe, although no eggs or young are ever found, provoked speculation from the earliest times. The ancient Greek philosopher Aristotle even suggested that the eels generated spontaneously in the mud at the bottom of lakes.

Later it was realized in northern Europe, in the countries fringing the Mediterranean Sea, and along the eastern seaboard of North America, that at certain times of the year small larval eels, or elvers, arrived in river estuaries. They then moved upriver, into inland freshwaters, where they proceeded to grow but never to breed. At another season, larger eels, their skin now silver

and their eyes much bigger appeared to move back into estuarine waters. Although these facts were known, they brought no closer a solution of the central, teasing question of where the adult eels actually bred.

It was not until the first decades of this century that the mystery was largely laid bare by the painstaking scientific detective work of the Danish oceanographer Johannes Schmidt (1877–1933). Near the end of the 1800s, it was determined that the tiny, elongate, leaf-shaped fish, known as leptocephali, which were sometimes caught in the open waters of the Atlantic, were not a new type of fish. Although quite different in appearance from the familiar elvers, and the glass eel stage that precedes them, they were shown to be young eel larvae. But where had they come from?

Starting in 1904, off the Faroe Islands well to the north of Scotland, Schmidt hunted these crucial, transparent slivers among the plankton. He found that the farther south and west he went, the more leptocephali he found, and the smaller they were. It seemed that he was approaching the place where they had been spawned. Eventually the breeding place of the eels was discovered—in the Sargasso Sea, at a depth of 300–600m/ 1,000–2,000ft beneath the protective

canopy of the sargassum weed. Here, too, the entirely marine conger eel, which grows even bigger than its freshwater cousin, has its breeding ground.

The fact that they breed in this remote area implies an immensely long two-way migration for the eels. Once hatched, the larvae drift with the currents, either back up the east coast of North America, or across to Europe with the Gulf Stream on a migration that takes several years. After some 10 years living and growing in freshwater, the adults return to the sea and again embark on the long journey to breed, and die, in the Sargasso.

A residual mystery clings to the eels' life history, however. Some scientists, although accepting the evidence for the larval migration, believe there is no direct proof that the reverse migration of European adult eels takes place. They think it is the North American eels, with their far shorter trek to and from the Sea, that provide the breeding pool for Europe's freshwater eels as well. If this is so, the European fish have only a one-way ticket when they leave the Sargasso and will never breed. This seems inherently unlikely, yet the eel's life is so extraordinary in other ways it would be a brave scientist who maintained this final twist was completely impossible.

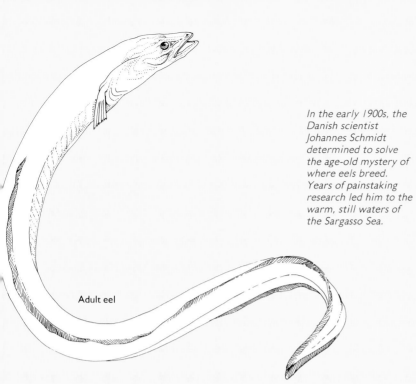

Adult eel

In the early 1900s, the Danish scientist Johannes Schmidt determined to solve the age-old mystery of where eels breed. Years of painstaking research led him to the warm, still waters of the Sargasso Sea.

Young adults of the species Anguilla anguilla, *the European eel, appear in their grey-green and brown livery in rivers throughout Europe, and as far east as Egypt and even the Black Sea. They change into their more familiar black and silver coloration only when they are fully mature and ready to make the journey back to the Sargasso Sea to breed.*

Dwellers in a watery wilderness

The Marshlands of southern Iraq are one of the world's last great secret places, where people have lived in much the same way for the past 6,000 years. The largest area of wetlands in the world, the marshes lie between the lower stretches of the Tigris and Euphrates rivers, some 225km/140mls southeast of Baghdad and 64km/40mls northwest of Basra, Iraq's major seaport.

The exact area is hard to define—it is ephemeral, moving constantly with the flooding of the two great rivers in spring and early summer, to create a vast sea where before there was desert. At the height of the flood season, in May and June, most of the region from Basra to Kut on the Tigris—some 322km/200mls—can be underwater, with only the higher parts of the land visible as islands.

Both the Tigris and Euphrates rise in the mountains of Turkey, thundering through deep canyons and narrow

An air of mystery, almost unreality, envelops the Marshland villages, paradoxically cut off by vast sheets of water from the desert world around them. Groups of small man-made islands dot the wilderness of reeds and water, each supporting a house constructed of bundled reeds, with closely woven reed mats unfurled to make the roof, walls and floor.

The islands' presence is revealed by smoke rising from fires of dried buffalo dung and by the passage of many small canoes. Occasionally a large, flat-bottomed barge is glimpsed, transporting merchandise to the edge of the marshes.

The area shaded yellow on the map shows the location of the Marshlands. It is reproduced on a larger scale on page 221.

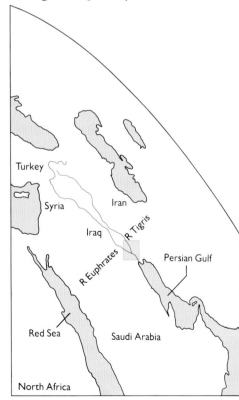

Turkey
Syria
Iran
Iraq
R Tigris
R Euphrates
Persian Gulf
Red Sea
Saudi Arabia
North Africa

gorges, and carrying along great loads of silt and alluvium stripped from the land. These clay deposits are dumped as the rivers reach the flatter plains of Iraq and Syria respectively, turning parched Earth into fertile agricultural land.

Their energy spent, the rivers run roughly parallel through the ancient valley of Mesopotamia and the flatlands of Sumer, their muddy, sluggish waters mixing together in places, through shifting natural channels and ancient canals. At Qurnah, 64km/40mls north of Basra, they meet at last in the area thought by some to be the site of the Garden of Eden. From here they flow as the Shatt al Arab through a vast delta to the Persian Gulf.

The Marshlands have always existed, but in times past they were not as extensive as they are today. In about 3500BC, the waters of the Tigris and Euphrates were tamed in Mesopotamia by the Sumerians, who constructed a sophisticated and skilful water-control system. Great tracts of land were reclaimed from the marshes, irrigation canals were built and maintained, dykes enclosed whole towns and villages set below sea level. The Babylonians further developed the water system. But except for a brief period in the eighth century, when once again the marshes were reclaimed, the Tigris and Euphrates have been overflowing their lower banks in times of flood, extending the Marshlands to what we see today.

The lifestyle of the Ma'dan, the Marsh Arabs, has changed little over the centuries. Once inside the forest of *qasab*— the 6-m/20-ft reeds that grow in the marshes—the world seems soundproofed and remote. The water is clean and calm and not very deep (averaging about 2.4m/8ft). Here, small canoes, called *mashhufs*, are the only way to travel; every family has at least one.

The hardy craft of the Ma'dan have been built in the same way since the Sumerians first paddled through the reedbeds in search of wildfowl and fish. But since no trees grow in the area, the boats are now made from timber imported from Malaysia and Indonesia, then waterproofed with bitumen that

bubbles from natural springs at Hit and Ramadi, in central Iraq.

The villages of the Ma'dan, scattered throughout the marshes, are also right out of history. The Babylonian story of the Creation, recorded in 2000BC, tells how Marduk, father of the gods, made the heavens. Later, "He built a reed platform on the surface of the waters, then created dust and poured it around the platform." So was the world created. And so, too, do the Ma'dan today create their own personal little worlds; for before a Ma'dan builds his home, he must still first build his island.

The extraordinary process consists of heaping great quantities of cut rushes in the water inside a sunken reed fence, sometimes spreading layers of mud between the rushes, and then trampling the whole mass down. When the heap reaches the surface of the water, the reed fence is folded in on top of it, and more reeds are trampled down to make a compact island, which seems to last for ever. Some of the deserted islands probably date back thousands of years.

The newly made island must be big enough to accommodate not only the family house, but also the water buffalo, cattle and sheep, which spend each summer night and all winter on the platform outside the house. Fodder, in the form of reed shoots, sedges and

Life in the marshes has a timeless quality, imposed by the yearly flooding of the great rivers, and ancient traditions persist. The Ma'dan fish as their distant ancestors did, standing in the prows of their canoes with five-pronged tridents at the ready.

The life and world of these strong, independent people was revealed to the world in the 1950s by the books and pictures of the last great explorer in Arabia, Wilfred Thesiger.

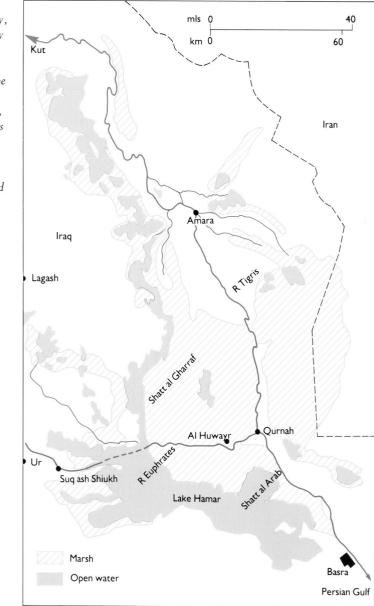

bulrushes, is provided for them daily.

Natural islands of packed mud do exist, most of them about 3m/10ft above water level. Thought to be the sites of ancient cities inhabited by spirits of the past, they are mysterious places in which the Ma'dan bury their dead.

Today, jet planes fly overhead; motor launches speed down the wider waterways; fishermen catch fish with nets; doctors and teachers visit the remote villages; and government aid allows rice cultivators to buy dehusking and winnowing machinery. The future of the timeless world of the marshes is in the balance; the Ma'dan are now poised on the edge of the sea of modernity.

The guesthouse, or mudhif, in each village is an imposing structure, usually 15.2m/50ft or more long and 4.6m/15ft wide and high, with a great vaulted roof supported on massive columns of bundled reeds. It is used for social gatherings and for entertaining visitors with strong, bitter coffee, rice, chicken and fish.

Wanderers of the Atlantic Ocean

If you were floating in the ocean a thousand miles from the nearest continental shore, could you detect the presence of an island beyond the horizon? Do islands somehow radiate non-visual clues to their existence? Human beings do not have the ability to recognize such clues, but the remarkable Atlantic green turtle, *Chelonia mydas*, apparently can sense them.

Every two or three years green turtles, living off the coast of Brazil where they graze on sea grass, swim to Ascension Island in the middle of the South Atlantic to mate and lay their eggs in the same nesting area, often on the same beach, as they themselves hatched. The island, a tiny speck in the ocean 8km/5mls wide, is 2,253km/1,400mls away, yet somehow they have developed the ability to find it.

What route do they follow? What guides the turtles; and what makes them migrate in the first place? Certainly, ocean currents in the Atlantic could determine their route, but it is difficult to believe that this is what actually happens. First, ocean currents are not constant, and they are frequently interrupted by winds and other currents. Second, it would take a long time to follow the roundabout routes of the currents, and where would the herbivorous turtles find food? Third, the waters of ocean currents are often extremely cold, particularly those of the southerly West Wind Drift, whose

Strong flippers working in unison like the wings of a bird, great green turtles roam half-way across the South Atlantic to breed on Ascension Island. What urges them to make this amazing journey and how they find their way is a mystery, but their brain structure suggests that they have an acute sense of smell and an inbuilt compass sense, both of which may help to guide them back to the beaches of their own hatching.

Green turtles may swim north or south along the Brazilian coast to make use of ocean currents in their migration, although it is more likely that the adults take a direct route, against the flow of the South Equatorial Current.

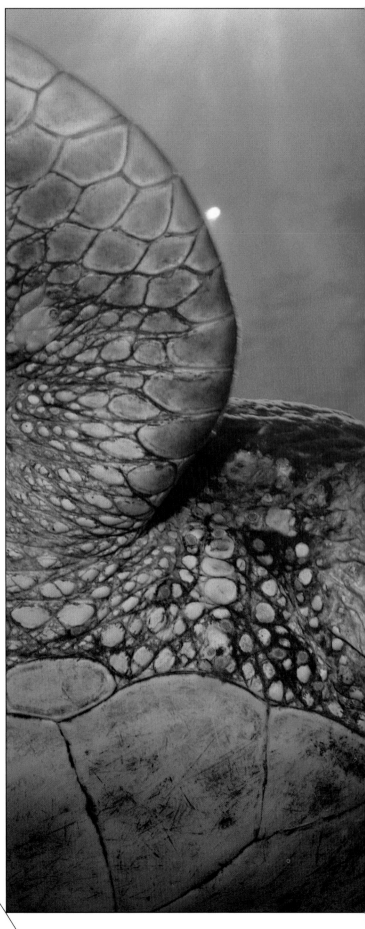

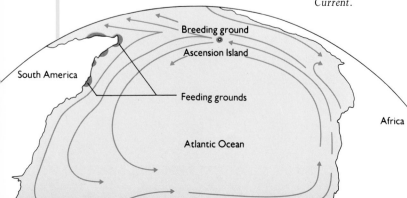

South America

Breeding ground

Ascension Island

Feeding grounds

Atlantic Ocean

Africa

temperature can be as low as 4.4°C/40°F, which would be lethal for turtles.

The most likely route for the mature turtles to take is probably the direct easterly one against the west-flowing South Equatorial Current. In doing this they would conform to the classic pattern of migration for aquatic animals: the strong adult animals move upstream, and the weak, inexperienced young swim downstream.

Presupposing that the hatchlings take with them some imprinted memory of the smell or taste of Ascension water when they take to the sea, this route would enable the returning adults to detect any effusions from the island quite far downstream in the South Equatorial Current. They could then home in on this smell or taste until they actually see the island. Quite probably this would be from some 32km/20mls away, since Green Mountain reaches 1,524m/5,000ft above sea level and is often capped by clouds rising well above this height. Other visual evidence might be seabirds converging on the island.

Assuming that Ascension Island does emit something that green turtles can detect, it is still necessary to explain how the turtles navigate across open ocean to reach it. One theory is that the turtle has some organ in its body enabling it to navigate by the stars. Another, more likely, is that the animal has a compass sense which enables it to judge the height of the Sun at noon and hence to estimate its latitude.

The compass theory proposes that green turtles travel north or south along the coast of Brazil, guided by their sense of smell or taste, until they arrive at the point of their first landfall as hatchlings. The latitude of this point would be close to that of Ascension Island. The turtles then swim east against the South Equatorial Current, their compass sense ensuring that they keep to the same latitude, until they approach the zone where their other senses pick up the mysterious emissions from the island. But this theory does not explain how the green turtle makes corrections to its compass course to allow for drift due to winds or ocean current fluctuations.

After two to three months, the hatchlings dig their way to the surface and at once set off for the sea, which they seem to find by the quality of the light over the water; the males never again return to land. Few of the young survive, for they are preyed on by other turtle species, fish and seabirds, while they themselves feed on small invertebrates, jellyfish, crabs and sea urchins. But by the time they are a year old, they have adopted the adult diet of sea grass.

With her hind flippers, the female turtle digs a 40-cm/16-in hole directly under her tail, in which she lays some 100 5-cm/2-in round, white, soft-shelled eggs. She then scoops sand over them and lumbers back to the ocean.

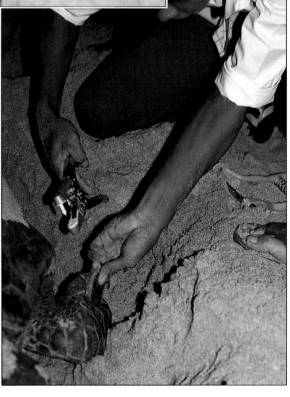

By tagging turtles, usually on the front flipper, scientists hope to track their movements. The best time for tagging is just after the females have laid their eggs, when they seem less afraid of humans.

To test any theory requires tracking a mature turtle over its entire migration, but so far this has not been done. For short distances, turtles towing floats, to which helium-filled balloons are attached as position markers, have been used by the University of Florida; for longer distances, radio transmitters have been mounted on turtles' backs. A turtle could, however, easily tow a raft on which has been mounted a transmitter and its power source, so that the radio

The green turtles may have colonized Ascension Island by accident in the first place, with egg-bearing females being carried there from the shores of West Africa, for West African ridley turtles seem to have reached coasts north of Brazil in this way. If green turtles had landed by chance on Ascension and nested, selective evolution would have ensured that the hatchlings would be imprinted with information enabling them to return to the island as adults, to breed. But the mystery remains. What can be the survival value of a colony of green turtles nesting on a remote mid-oceanic island, when to do so requires delicate sense organs not found in any other turtle?

Oasis of luxuriant life

In the interior of southern Africa, in Botswana, lies one of creation's great mysteries. Here the waters of the Okavango, the region's third largest river, fan out over the sands of the Kalahari Desert. In a maze of criss-crossing channels, swamps, islands and floodplains they form the largest inland delta on Earth, where water behaves in a most peculiar fashion.

The waters of the Okavango Delta support a diverse ecosystem, rich in plant, animal, fish and bird life. Humans, too, have utilized its rich potential for hunting for tens of thousands of years, although settlement within the delta has always been limited. But the outside world knew nothing of the Okavango until 1853, when Charles Andersson, (1827–67), a Swedish explorer, became the first European to set eyes on it.

He journeyed through part of the system of channels, which he considered to be "of an indescribably beautiful appearance". Not only is the delta beautiful, it is deeply mysterious: an oasis in the semi-arid Kalahari; the end-point of a major river that flows inland, rather than to the sea; a region where great volumes of water flood in each year, then simply disappear.

The delta region lies some 1,000km/ 621mls from the South Atlantic Ocean to the west and 1,300km/808mls from the Indian Ocean in the east, and covers around 15,000sq km/5,805sq mls. At its greatest length, the delta is more than 260km/162mls long, while the Oka-vango River extends a further 1,100km/ 684mls back to its source in the wet Angolan Highlands, where it is known as the Cubango River.

In an average year, this river and its largest tributary, the Cuando, bring 1,050 million cu m/37,065 million cu ft of water to the delta; local rains add another 500 million cu m/17,650 million cu ft. One of the peculiarities of

The Okavango Delta, the largest and most beautiful oasis in Africa, is one of Earth's strangest phenomena. For here, in the middle of the burning Kalahari sands, the great river spreads out over an area almost the size of Wales and then vanishes.

During the floods, the reedy channels and peaceful lagoons nurture an astonishing diversity of living creatures. Even the decorative water lilies provide pollen for bees and seeds for pygmy geese, while within the flower cups midges lay their eggs.

the delta is, however, that only some 2–5 percent of the total amount of water leaves it through the periodic rivers that flow from this widespread system of swamps and channels.

The clearest overall view of the topography of the region is gained from the air and reveals that the main part of the delta is roughly triangular, with the apex in the northwest near the village of Seronga. The base of the triangle runs southwest to northeast and is very abrupt, particularly along the stretch that extends northeast from Maun.

Although the Okavango River reaches the "triangle" at Seronga, it has already been in the delta zone for some 90km/56mls, passing through a section known as the "Panhandle", whose shape is determined by geological faults. In this stretch, the course of the river itself—up to 6m/20ft deep and 150m/490ft wide—is still identifiable, but it becomes progressively more broken by reedy swamps and small sandy islands, colonized by trees and grasses.

At the apex of the triangle, the waters begin to fan out into the delta proper, with the Okavango River finally splitting up into a system of major channels, four of which—the Thoage, Mboroga, Jao and Maunachira—weave their way slowly to the triangle's base. These channels may peter out altogether or merge again to form larger streams which skirt peat and sand islands, such as the 600-sq km/230-sq ml Chief's Island in the heart of the delta.

The fifth distributary, the Selinda Spillway, or Magwegqana, follows a different pattern. No sooner has it entered the delta than it leaves again, heading toward a second but much smaller delta, the Linyanti Swamp, some 120km/75mls to the east. The pattern of the delta from the air has led one scientist to suggest that it looks like a giant hand: the Panhandle is the wrist or sleeve and the Selinda Spillway the thumb, with the other major channels forming the fingers.

The most baffling aspect of the delta has always been why such a feature should exist in the heart of Africa, and why a large river such as the Okavango

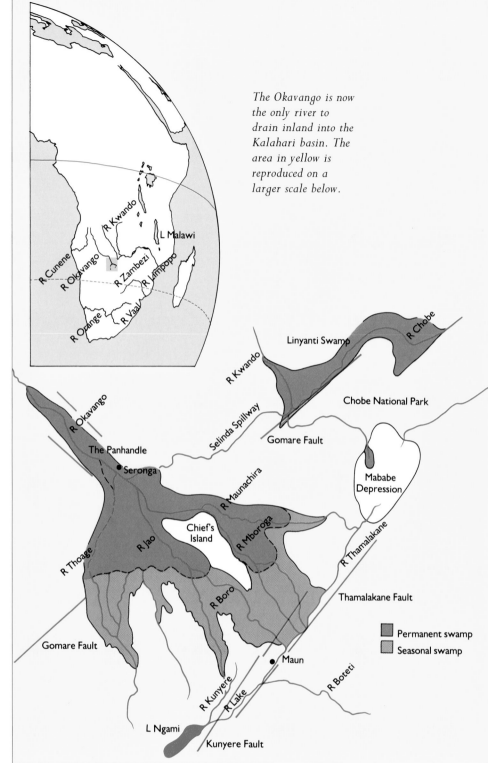

The Okavango is now the only river to drain inland into the Kalahari basin. The area in yellow is reproduced on a larger scale below.

should lose its water inland, rather than into the ocean. Geological research over the past 20 years has provided many of the answers, though one perceptive geologist, the South African Alex du Toit (1878–1948) had identified major components of the explanation as long ago as the 1920s.

The African continent has existed for about 80 million years. Before that, Africa formed the central part of a much larger landmass, the supercontinent of Gondwana. Around 200 million years ago, Gondwana began to break up under the effect of Earth movements. Antarctica broke away from Africa's east coast about 180 million years ago, and the west coast began to take shape some 130 million years ago, as South America and Africa drifted apart.

The division of Gondwana did not leave the central land area, now southern Africa, unscathed. The rifting that marked the formation of the surrounding oceans forced the land some 100–300km/62–186mls from the coast up to

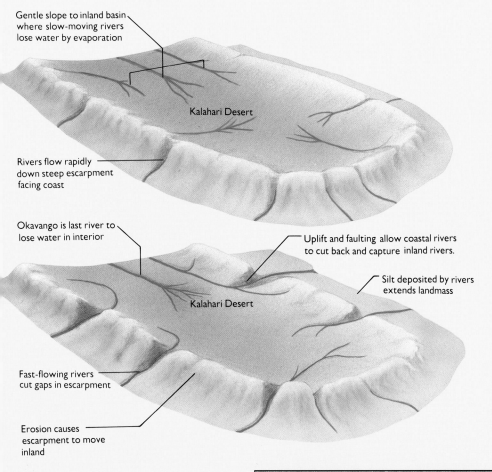

Gentle slope to inland basin
where slow-moving rivers
lose water by evaporation

Kalahari Desert

Rivers flow rapidly
down steep escarpment
facing coast

Okavango is last river to
lose water in interior

Uplift and faulting allow coastal rivers
to cut back and capture inland rivers.

Silt deposited by rivers
extends landmass

Kalahari Desert

Fast-flowing rivers
cut gaps in escarpment

Erosion causes
escarpment to move
inland

How the delta was created

The Great Escarpment divided the rivers of southern Africa into two systems: relatively short, steep rivers, draining into the sea, and rivers with gentle gradients flowing into the great interior Kalahari basin.

Evidence from the landscape indicates that some major rivers, including the Cuando and the Zambezi, whose upper courses parallel that of the Okavango, once flowed into the interior and formed part of the inland drainage system. Except for the Okavango, these have been captured by the more aggressive coastal rivers, which have cut back through the escarpment to tap the ancient inward-draining streams.

The delta itself has developed because the Okavango River flows into an area underlain by a geological rift structure. This rift runs northeast to southwest, with its southeastern side, the Thamalakane Fault, forming the downstream limit of the delta. Unlike the Great Rift Valley, of which it is the southwestern extension, it does not have steep valley sides, since river sediments have, over the millennia, filled it in to a thickness of 300–1,000m/980–3,300ft.

A satellite picture of the Okavango Delta shows clearly how the rift has ponded back the river. As a result, its waters have fanned out into smaller channels, and its load of sediment has been deposited.

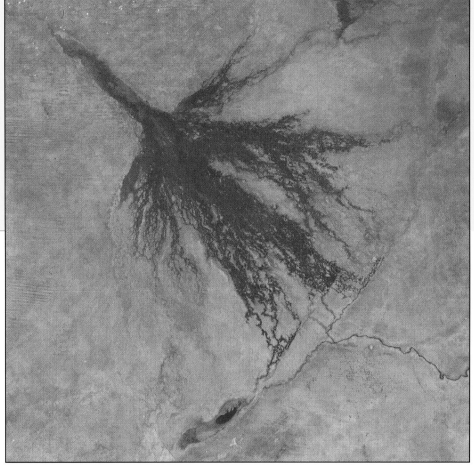

form a distinct ridge of higher ground. Known as the Great Escarpment, this separates the coastal plains from the interior, which has itself undergone gentle downwarping, so creating a shallow high-altitude basin on average 1,000m/3,300ft above present sea level. Into this basin flows the Okavango River, to lose itself in the Kalahari sands.

The northern part of the delta receives water from the river all year round; the central and southern parts experience an annual cycle of events which replenishes the waters of the channels and swamps. The yearly flood that follows the summer (November-February) wet season in Angola does not reach the delta until April, and takes a further three months to reach Maun. As the flood arrives, water levels in the papyrus-fringed channels rise by around 1m/3ft—sufficient to inundate the grassy, palm-dotted plains that border the wooded delta islands and the smaller seasonal channels.

As the floodwater fans out into the delta, its surface area increases. This makes it more susceptible to evaporation, the main cause of water loss, especially since the arrival of the flood coincides with the dry season. The air above the delta is warm and extremely dry and so is able to pick up a great deal of moisture. Some water does, however, flow out of the delta, southwest along the Thamalakane Fault toward the now dry Lake Ngami and northeast to the Mababe Depression. A little water escapes completely from the system, down the Boteti River and into the Kalahari.

By September, flood levels within the delta are usually falling rapidly, but this does not mark the end of the year's water supply. In November, the local rains arrive and last until March; they bring on average some 30–35 percent of the delta's water. The total amount of water received can vary markedly from year to year, depending on rainfall both in the catchment area and directly over the delta. In any one year, distributary channels can carry water through less than half of the delta area. But the distribution of alluvial, or riverborne, sediments in areas fringing the modern system of channels indicates that, on at least one occasion in the geological past, conditions must have been wetter than they are today, for they cover a further 7,000sq km/2,700sq mls.

The swamps and fertile plains of the Okavango Delta, astounding in the middle of such a dry, dusty expanse, support a rich and diverse wildlife. Here species adapted to wet environments meet with those common in the Kalahari Desert. The animal life varies both

seasonally, according to the amount of water, and in response to the different ecosystems which the delta contains.

More than 400 species of bird have been identified, many of them, such as the pink-backed pelican, the wattled crane and the slaty egret, rare. Indeed, the delta is the only place in the world where slaty egrets are known to breed. In the channels and lagoons 65 fish species are found—food for birds and humans alike. And the channels are also home to hippopotamuses and crocodiles, while the lagoons and flood plains support two species of antelope, the red lechwe and the sitatunga, which are specially adapted to life in a water environment.

The larger islands have resident populations of elephant and buffalo. Giraffe, zebra and many antelope species found in the Kalahari, together with the predatory lion, leopard, cheetah, hyena, and Cape hunting dog, all gather in the delta

and on its fringes to take advantage of the bounty of the flood: fresh grass, foliage and water.

In recent centuries, various Tswana groups have been the main human occupants of the delta area. The difficulties of travel and the dangers of contracting malaria and sleeping sickness have, until a few years ago, restricted the occupation and utilization of the delta to its fringes. One group, the Ba Yei, are traditionally regarded as hunters and fishermen, moving about the channels in dug-out canoes; while another, the Hambukushu, have practised small-scale farming on some of the delta's islands.

The last few decades have brought a range of pressures to bear upon the delta and its inhabitants. Botswana is a dry country, most of it desert, and the Okavango is its largest supply of surface water. It is not surprising, therefore, that various schemes have been talked of, but none so far implemented, to tap

The flooded channels of the delta are full of fish, which the Ba Yei women catch in traditional baskets made from reeds.

A dust cloud, visible for miles, marks the migration route of thousands of buffalo, Synceros caffer, into the delta at the end of the dry season. Other huge herds stay in the area year-round.

Herds of Hippo, Hippopotamus amphibius, live permanently in the lagoons. During the day they loll in the water, almost invisible, emerging occasionally to sun themselves. At night they come ashore to feed. Their trampling makes paths through the dense reedbeds, helping to keep open channels in the swamps.

The tribesmen of the delta are expert river people, who use their dug-out canoes, or mokoros, *for hunting and fishing. Their way of life is in harmony with the ecology of the area, and they do little to harm it.*

its waters and divert them to areas farther south, where there is a large and growing demand for water from expanding human and livestock populations.

Within the delta itself, wildlife is under pressure, though the sources responsible for this have changed during the past 30 years. In 1956–68, for example, commercial crocodile hunting accounted for more than 40,000 animals, a depletion from which the population is now recovering. The extension of human settlement into the western fringes of the delta has been injurious to wildlife in general, and fish and hippopotomuses in particular, but the establishment of the 3,800sq-km/1,470sq-ml Moremi Wildlife Reserve in the east has provided a protected and favourable environment for the indigenous animals. Together with the adjacent Chobe National Park, this reserve today contains some of Africa's most abundant and spectacular wildlife.

The abundance of wildlife brings its own pressures, however, and since the 1970s the Okavango Delta has become an important tourist attraction. Many camps have been established on islands often accessible only by light aircraft or boat. The visitors' presence, especially if numbers continue to grow, is becoming a threat to the very "wilderness experience" they seek.

Wildlife is also threatened by the expansion of domestic livestock grazing into the delta. But the construction of a buffalo-proof fence around part of the delta margin in the 1960s, to prevent the spread of foot-and-mouth disease to cattle, resulted in benefit to the wildlife as well, by keeping the cattle out. The drought that hit Botswana during the 1980s created renewed demands for a greater expansion of cattle rangelands into the delta. This has been made all the more possible by the virtual eradication of the tsetse fly, which spreads sleeping sickness to domestic livestock.

Perhaps the conflicting demands of humans, livestock and wildlife will be partially resolved by the recent exploitation of the world's second richest source of diamonds at Orapa, not far from the Okavango Delta. Diamond mining has now become Botswana's most prosperous industry, followed in fourth place by tourism, which may help lessen people's dependence on their cattle for a living. This may in turn relieve the pressure on the delta— unless the mines' demands for water simply create a different threat.

The Okavango Delta is, of itself, a dynamic ecological system with a unique web of interdependencies. But without careful management, it seems likely that this strange and magnificent wetland, with its breathtaking variety of free-roaming wildlife, will gradually be taken over and its character destroyed. So, eventually, a natural mystery may be lost to the world. The struggle in Botswana to maintain the fragile balance in the face of conflicting needs is a microcosm of the greatest problems facing all countries and all peoples on planet Earth as we approach the twenty-first century.

Tourism has brought a stream of
visitors to the delta. Unchecked, their
numbers, and pollution, may destroy
the beauties the tourists seek; but the
cash they bring could also be an
incentive for the conservation of both
wildlife and habitat.

Bibliography

This list comprises a selection of those books consulted by the Publishers in the preparation of this volume and also suggestions for further reading.

Allen, O.E. *Planet Earth: Atmosphere* Time-Life Books, Amsterdam, 1983

Andersson, C.J. *Lake Ngami or Explorations and Discoveries During Four Years' Wanderings in the Wilds of South Western Africa* Hurst and Blackett, London, 1856; facsimile reprint C. Struik, Cape Town, South Africa, 1987

Antarctica: Great Stories from the Frozen Continent Reader's Digest, Sydney, London & New York, 1985

Atkinson, B.W. & Gadd, A. *Weather: A Modern Guide to Forecasting* Mitchell Beazley, London, 1986

Baker, R. *Migration Paths Through Time and Space* Hodder & Stoughton, London, 1982; *The Mystery of Migration* (Ed) Macdonald, London, 1980

Banyard, P.J. & Black, D. *Wonders of the World* Orbis Publishing, London, 1984

Bárdárson, H.J. *Ice and Fire* H.R. Bárdárson, Reykjavik, Iceland, 1971

Bere, R.M. *The Way to the Mountains of the Moon* Arthur Barker, London, 1966; *The Wild Mammals of Uganda* Longmans, London, 1962

Bond, C. *Okavango: Africa's Last Eden* Country Life Books/Hamlyn, London, 1978

Botting, D. *Humboldt and the Cosmos* Michael Joseph, London, 1973

Boy, G. & Allan, I. *Snowcaps on the Equator* The Bodley Head, London, 1989

Boyle, S. & Ardill, J. *The Greenhouse Effect* Hodder & Stoughton, London, 1989

Branston, B. *The Last Great Journey on Earth* Hodder & Stoughton, London, 1970

Cloudsley-Thompson, J. *Animal Migration* Orbis, London, 1978

Cox, C.B. & Moore, Peter D. *Biogeography* (4th Edition) Blackwell Scientific Publications, Oxford, UK, 1985

Cox, B., Savage, R.J.G., Gardiner, B. & Dixon, D. *Macmillan Encyclopedia of Dinosaurs and Prehistoric Animals* Macmillan Publishing, New York and London, 1988

Dudley, W. & Lee, M. *Tsunami!* University of Hawaii Press, Honolulu, Hawaii, 1988

Ehrlich, P. & A. *Extinction: The Causes and Consequences of the Disappearance of Species* Random House, New York, 1981

Erikson, J. *Violent Storms* Tab Books, Blue Ridge Summit, PA, USA, 1988

Falck-Ytter, H. *The Aurora: The Northern Lights in Mythology, History and Science* (Trans R. Alexander), Floris, Edinburgh, UK, 1985

Filippi, F. de *Ruwenzori: An Account of the Expedition of HRH Prince Luigi Amadeo of Savoy, Duke of the Abruzzi* Archibald Constable, London, 1908

Fisher, R., Gray, W.R., McIntyre, L., O'Neill, T. and Ramsay, C.R. *Nature's World of Wonders* National Geographic Society, Washington, D.C., USA, 1983

Gendron, V. *The Dragon Tree: A Life of Alexander, Baron von Humboldt* Longmans, Green, London and New York, 1961

Graves, N., Lidstone, J. & Naish, M. *People and Environment: A World Perspective* Heinemann Educational Books, London, 1987

Grelier, J. *To the Source of the Orinoco* Herbert Jenkins, London, 1957

Gribbin, J. *The Hole in the Sky: Man's Threat to the Ozone Layer* Corgi Books, Transworld Publishers, London, 1988; *This Shaking Earth* Sidgwick & Jackson, London, 1978

Gribbin, J. & Kelly, M. *Winds of Change: Living in the Global Greenhouse* Hodder & Stoughton, London, 1989

Gudmundsson, A.T. & Kjartansson, H. *Guide to the Geology of Iceland* Bókaútgáfan Örn Og Örlygur HF, Reykjavik, Iceland, 1984

Halstead, L.B. *Hunting the Past* Hamish Hamilton, London, 1982

Hanbury-Tenison, R. *The Rough and the Smooth: The Story of Two Journeys Across South America* Robert Hale, London, 1969

Hardy, R. Wright, P. Gribbin, J. & Kington, J. *The Weather Book* Michael Joseph, London, 1982

Holford, I. *The Guiness Book of Weather Facts and Figures* Guiness Superlatives, Enfield, Middlesex, UK, 1982

Jackson, D.D. *Planet Earth: Underground Worlds* Time-Life Books, Amsterdam, 1982

Jackson, M.H. & Lopez, B. *Galapagos* Rizzoli, New York, 1989

Jacobs, M. *The Tropical Rain Forest: A First Encounter* (Trans. R. Kruk) Springer Verlag, Berlin, 1988

Kellner, L. *Alexander von Humboldt* Oxford University Press, London and New York, 1963

Lambert, D. & the Diagram Group *The Cambridge Guide to the Earth*, Cambridge University Press, Cambridge, UK, 1988

Leakey, R.E. *Human Origins* Hamish Hamilton, London, 1982; *The Making of Mankind* Michael Joseph, London, 1981

Lewis, T. (Ed) *Planet Earth: Volcano* Time-Life Books, Amsterdam, 1982

Lockhart, G. *The Weather Companion* John Wiley & Sons, New York, 1988

Lutgens, F.K. & Tarbuck, E.J. *Essentials of Geology* Charles E. Merril Publishing Company, Columbus, Ohio, 1986

Main, M. *Kalahari: Life's Variety in Dune and Delta* Southern Book Publishers, Cape Town, South Africa, 1987

May, J. *The Greenpeace Book of Antarctica* Dorling Kindersley, London, 1988

Michell, J. & Rickard, R.J.M. *Living Wonders: Mysteries and Curiosities of the Animal World* Thames & Hudson, London, 1982

Moore, P., Hunt, G., Nicolson, I., Cattermole, P. *The Atlas of the Solar System* Mitchell Beazley in Association with the Royal Astronomical Society, London, 1983

Nicolson, I., Moore, P. *The Universe* Collins, London, 1985

Moorehead, A. *Darwin and the Beagle* Hamish Hamilton, London, 1969

Palmer, R. *Deep Into the Blue Holes: The Story of the Andros Project* Unwin Hyman, London, 1989

Potgieter, H. & Walker, C. *Above Africa: Aerial Photography from the Okavango Swamplands* New Holland, London, 1989

Prance, G.T. & Lovejoy, T.E. *Amazonia (Key Environments)* Pergamon Press, Oxford, UK, 1985

Richards, P.W. *The Tropical Rain Forest* Cambridge University Press, Cambridge, UK, 1979

Ross, K. *Okavango: Jewel of the Kalahari* BBC Books, London, 1987

Rowbotham, F. *The Severn Bore* David & Charles, Newton Abbot, Devon, UK, 1970

Rubeli, K. *Tropical Rain Forests in South-East Asia: A Pictorial Journey* Tropical Press, Kuala Lumpur, Malaysia, 1986

Smith, P.J. (Ed) *Encyclopedia of the Earth* Hutchinson, London, 1986

Synge, P.M. *Mountains of the Moon* Lindsay Drummond, London, 1937

Thesiger, W. *The Marsh Arabs* Longmans Green, London, 1964

Tilman, H.W. *Snow on the Equator* G. Bell & Sons, London, 1937

Thomas, D.S.G. & Shaw, P. *The Kalahari Environment* Cambridge University Press, Cambridge, UK, 1990

Verney, P. *The Earth Quake Handbook*, Paddington Press, London & New York, 1930

Walker, B. *Planet Earth: Earthquake* Time-Life Books, Amsterdam, 1982

Webster, D. *Understanding Geology* Oliver & Boyd, Edinburgh, Scotland, 1987

Whipple, A.B.C. *Planet Earth: Storm* Time-Life Books, Amsterdam, 1982

Whitfield, P. *The Atlas of the Living World* Weidenfeld & Nicolson, London and Houghton Mifflin, New York, 1989

Whitmore, T.C. *Tropical Forests of the Far East* (3rd Edition) Clarendon Press, Oxford, UK, 1984

Wood, M. *Different Drums: Reflections on a Changing Africa* Century, London, 1987

Wood, R.M. *Earthquakes and Volcanoes* Mitchell Beazley, London, 1986

Yeoman, G. *Africa's Mountains of the Moon* Elm Tree Books, London, 1989

Young, G. *Return to the Marshes: Life with the Marsh Arabs of Iraq* Collins, London, 1977

Index

Note: Page numbers in **medium type** indicate that more than a few lines are devoted to the subject. Page numbers in *italics* refer to illustrations and their captions.

q. = quoted

Acknowledgments

ARTWORK CREDITS

Maps by Technical Art Services

Dave Ashby 12–13, 20–1, 27, 33, 36–7, 62–3, 105, 135

Lynn Bowers 151, 154–5, 176–7, 216

Malcolm Ellis 188 (colour)

Vana Haggerty 181, 188 (black and white)

Steve Holden 200–1

Steve Kirk 194–5

Paul Richardson 106–7, 111, 119

Ed Stuart 57, 74–5, 82–3, 86, 91, 126–7, 138, 144, 171, 229

PICTURE CREDITS

l = left *r* = right *t* = top *c* = center *b* = bottom

Front cover Zefa Picture Library
inset Tad Janocinski/The Image Bank
Back cover *t* V. Englebert/Zefa Picture Library
l NASA/Frank Lane Picture Agency
Orion Press/NHPA
r Gordon Garradd/Science Photo Library
b Library
6–7 P. Vauthey/Sigma/The John Hillelson Agency

9 Dr Jean Lorre/Science Photo Library
10–11 Zefa Picture Library
13 NASA/Science Photo Library
14*t* Tony Morrison/South American Pictures
b, 15 Michael Holford
16–17 NASA/Science Photo Library
17*t* Ann Ronan Picture Library
b Vivien Fifield
18 Martyn Chillmaid/Oxford Scientific Films
18–19 John Downer
20 The Mansell Collection

22*l* & *r*	Michael Holford/Science Museum, London	90–1	Robert Hessler/Planet Earth Pictures	157*t*	Aldus Archive
23*t*	Zefa Picture Library	*l* & *r*		*b*	John Cleare/Mountain Camera
b	National Maritime Museum, London/ Bridgeman Art Library	91	Peter Ryan/Science Photo Library	158–9,	Richard Packwood/Oxford Scientific
24	Ned Haines/Science Photo Library	92	John Noble/Mountain Camera	159	Films
24–5	Jack Finch/Science Photo Library	92–3	Colin Monteath/Mountain Camera	160–1	Gerald Cubitt/Bruce Coleman
26*t*	University Library, Oslo	94	Doug Allan/Science Photo Library	162–3	Peter Scoones/Planet Earth Pictures
b	Baader Planetarium Archive	96	The Mansell Collection	163*l*	Ken Lucas/Planet Earth Pictures
28*t*	Germanisches Museum, Nürnberg	96–7	Popperfoto	*r*	FAO
b	Vivien Fifield	98	Doug Allan/Science Photo Library	164,	Douglas Botting
29	Jack Finch/Science Photo Library	98–9	Colin Monteath/Mountain Camera	164–5,	
30–1	Messerschmit/Zefa Picture Library	99	Edwin Mickleburgh/Ardea	167	
32	Vivien Fifield	101	NASA/Science Photo Library	168–9	Anthony Suau/Black Star/Colorific!
33	NASA/Science Photo Library	102–3	Stockmarket/Zefa Picture Library	170, 171	P. Turnley/Black Star/Colorific!
34–5	Zefa Picture Library	104	NASA/Science Photo Library	172,	Antony Suau/Black Star/Colorific!
36–7	Link Observatory	105	The Mansell Collection	172–3	
38	F.W. Rowbotham	106*l* & *r*	R.K. Pilsbury	174–5	NASA/Science Photo Library
40–1	John Lythgoe/Planet Earth Pictures	107*tl*	John Cleare/Mountain Camera	176, 177*t*	Gavin Newman
41	Martin Dohrn/Science Photo Library	*tr*	Jean Hosking/Frank Lane Picture Agency	177*b*	Rob Palmer
42	NASA/Science Photo Library			178–9	Jerry Wooldridge/Planet Earth
42–3	François Gohier/Ardea	*c*	David W. Hamilton/The Image Bank		Pictures
44	NASA/Science Photo Library	*b*	J. Pfaff/Zefa Picture Library	180	Steenmanns/Zefa Picture Library
44–5	Photri/Robert Harding Picture Library	108–9	Sygma/The John Hillelson Agency	181, 182*t*	Chris Howes
45	NASA/Science Photo Library	110–11	NASA/Frank Lane Picture Library	182*b*	Michael Holford
46	Scala, Milan	111	Robert Harding Picture Library	183	Chris Howes
47*l*	Max-Planck-Institut/David Parker/ Science Photo Library	112–13	British Museum	184–5	Peter Scoones/Planet Earth Pictures
		114	By Anthony Boccaccio © National Geographic Society	186–7	Zefa Picture Library
r	Vivien Fifield			189	Sinclair Stammers/Science Photo
48	B. Crader/Zefa Picture Library	115*t*	W Carlson/Frank Lane Picture Library		Library
48–9	Zefa Picture Library			190	Frank Schneidermeyer/Oxford
50*t*	Iain Roy	*b*	Derek Elsom		Scientific Films
b	Robin Scagell/Science Photo Library	116	Central African Museum, Tervuren/ Werner Forman Archive	190–1	Helmlinger/Zefa Picture Library
51	Wardene Weisser/Ardea			191	Harry Taylor/Oxford Scientific Films
53	David Wright/Oxford Scientific Films	116–17	Gordon Garradd/Science Photo Library	192–3	Colin Orthner/Tyrell Museum of Paleontology
54–5	Horst Munzig/Susan Griggs Agency				
58	Martin Land/Science Photo Library	118*t* & *b*	Ann Ronan Picture Library	196	Professor Walter Alvarez/Oxford
58–9	George Hall/Susan Griggs Agency	119	Mary Evans Picture Library		Scientific Films
60–1	Larry Pierce/The Image Bank	120	Steve Krongard/The Image Bank	196–7	Keith G. Cox/University of Oxford
63	Reflejo/Susan Griggs Agency	120–1	Robert Harding Picture Library	198–9	Anthony Bannister/NHPA
64, 65*t*	Professor Stewart Lowther/Science Photo Library	122–3	Walter Rawlings/Robert Harding Picture Library	201	John Reader/Science Photo Library
				202	M.P. Price/Bruce Coleman
65*b*	Alan Kearney/Tony Stone Associates	123*t*	John Tindale	203*t*	John Reader/Science Photo Library
66–7	Bridgeman Art Library	*b*	Zefa Picture Library	*b*	R.I.M. Campbell/Bruce Coleman
67*t*	Michael Holford/British Museum	124–5	C. Carvalho/Frank Lane Picture Library	204–5	A.P.L./Zefa Picture Library
b	Bridgeman Art Library			206	Popperfoto
68, 68–9	Martyn Chillmaid/Oxford Scientific Films	126	Tony Morrison/South American Pictures	206–7	Royal Geographical Society
				207	Hulton-Deutsch Collection
70	Tony Stone Associates	127*t*	Riobbu Newman/The Image Bank	208	Brian Cotes/Bruce Coleman
70–1	Orion Press/NHPA	*b*	Gamma/Frank Spooner Pictures	209	Robert Harding Picture Library
71	Robert Harding Picture Library	128–9	V. Englebert/Zefa Picture Library	210*t*	Michael Marten/Science Photo Library
72–3	Joffet/Sipa Press/Rex Features	130	J.R. Bracegirdle/Planet Earth Pictures	*b*	Dr Jeremy Burgess/Science Photo
75	Popperfoto	131	V. Englebert/Zefa Picture Library		Library
76	Earthquake Research Institute, University of Tokyo	132–3	Michael K. Nichols/Magnum	211*t*	Kim Taylor/Bruce Coleman
		134	Herve Collart/Gamma/Frank Spooner Pictures	*b*	K. Wothe/Bruce Coleman
77*t*	Tony Stone Associates			212–13	David Doubilet
b	Michael Holford/Robin Gwynn (BBC)	134–5	Chris Newton/Frank Lane Pictures	214–15	James Carmichael/NHPA
78	Popperfoto	136–7	Chip Hires/Gamma/Frank Spooner Pictures	215	Peter David/Planet Earth Pictures
78–9	Robert Harding Picture Library			217*t*	Hulton-Deutsch Collection
79	T. Campion/Sygma/The John Hillelson Agency	138–9	J.G. Paren/Science Photo Library	*b*	John Lythgoe/Planet Earth Pictures
		141	NASA/Bruce Coleman	218–19	Tor Eigeland/Susan Griggs Agency
80–1	Michael Holford/Victoria and Albert Museum	142–3	Lee Lyon/Bruce Coleman	220–1	Wilfred Thesiger, *Visions of a Nomad*
		144–5	V Englebert/Susan Griggs Agency	221	Southwell/Robert Harding Picture
82–3	The Mansell Collection	145	Carl Purcell/Colorific!		Library
83	Dieter & Mary Plage/Bruce Coleman	146–7	Sean T. Avery/Planet Earth Pictures	222–3	James D. Watt/Planet Earth Pictures
84–5	John Lythgoe/Planet Earth Pictures	147	Jonathan Scott/Planet Earth Pictures	224*t*	Soames Summerhays/Biofotos
87*t*	British Museum (Natural History)	148–9	Peter Scoones/Planet Earth Pictures	*b*	Rod Salm/Planet Earth Pictures
b	Institute of Oceanographic Science/ Oxford Scientific Films	150	Hulton-Deutsch Collection	225	Soames Summerhays/Biofotos
		150–1	Peter Scoones/Planet Earth Pictures	226–7	J.R. Bracegirdle/Planet Earth Pictures
88	James M. King/Planet Earth Pictures	152–3,	John Cleare/Mountain Camera	229	David S.G. Thomas
89*t*	Hulton Deutsch Collection	154*l*		230–1	Partridge Films/Oxford Scientific Films
b	Dick Clarke/Planet Earth Pictures	154*r*, 155*t*	Biofotos	231*t*	Tom Nebbia/Aspect Picture Library
		155*b*	Patti Murray/Oxford Scientific Films	*b*	Partridge Films/Oxford Scientific Films
		156–7	John Cleare/Mountain Camera	232	Lee Lyon/Bruce Coleman
				232–3	Richard Coomber/Planet Earth Pictures

The contributors

This book was to have been written by DR EDWARD T. STRINGER, FRGS, FRMetS, FRC, who died when he had delivered just three chapters. The publishers acknowledge his enthusiastic contribution to the concept and planning of the book and feel he would have admired the work of those who took his place.
Pages 24–29; 152–159; 222–225

DAVID BURNIE studied zoology and botany at the University of Bristol. After graduating, he spent some time working in the Scottish Highlands, eventually becoming a nature reserve ranger before moving south to work as a biologist. In 1979, he started a career in writing and editing books on natural sciences and technology. He has contributed to a number of magazines and has written several books on topics ranging from birdwatching to the workings of machines.
Pages 60–83; 92–99

PROFESSOR BARRY COX, Assistant Principal and Professor of Biological Sciences, King's College, University of London, studied zoology at Oxford University and gained a research degree on fossil reptiles at Cambridge University. His interest in biogeography resulted from his collections of fossil reptiles made on expeditions to Africa and South America. His student textbook *Biogeography*, written jointly with Dr Peter Moore, has become a standard work in Britain and in North and South America. He has recently continued his study of the prehistoric world at Stanford University, California.
Pages 54–59; 192–197

CAROLE C. DEVANEY, who studied at Trinity College, Dublin, is a marine zoologist and a writer and editor of books on the natural sciences. Keen on photography, she takes pictures of archaeological sites on land and marine life underwater. She lives and works in Dublin.

Pages 128–131; 164–167; 218–221

DR DEREK ELSOM is the Principal Lecturer in Geography at Oxford Polytechnic and Director of the Research Centre of the Tornado & Storm Research Organisation (TORRO) at Oxford. He has written more than 50 papers on air pollution and weather hazards such as tornadoes, hailstorms and lightning, and his textbook, *Atmospheric Pollution*, has become a standard student text in air pollution studies. He has also written many articles for magazines such as *New Scientist* and the *Geographical Magazine*.

Pages 102–127; 132–139

DR LYNN FROSTICK, a geologist specializing in sedimentology, is an expert on desert environments. She spends a great deal of her time deciphering the secrets of past events, now locked in the rock record, by analogy with modern settings. Her work has involved expeditions to East Africa and the Levant, where desert sediments are preserved in the Great Rift Valley. She is a Senior Lecturer in Geology at the University of Reading.

DR IAN REID is Reader in (Associate Professor of) Geography, Birkbeck College, University of London. He specializes in hydrology, geomorphology and sedimentology and has a particular interest in the way rivers lay down sediments in rift valleys. His work has taken him to East Africa, Egypt and Israel, among other places, where he has collaborated with paleoanthropologists and archaeologists in the attempt to unravel the mysteries of our early ancestors.

Co-authors pages 142–151; 198–203

PETER L. SMART studied at the University of Bristol and the University of Alberta, Canada; he is at present Lecturer in Geography at Bristol. His chief interest is in the hydrology and geomorphology of limestone areas, and especially of caves. He is currently working on a major research project, on the hydrology and geochemistry of groundwaters in the Bahamas and the dating of past sea levels, which involves extensive diving in and exploration of "blue holes".

Pages 174–183

DR DAVID S.G. THOMAS was educated at Oxford University and has been Lecturer in Geography at the University of Sheffield since 1984. He is an expert on arid environments and has carried out research and other investigations in deserts in North America, the Near East and Africa, particularly in the Kalahari. He has written more than 30 scientific papers and two books, including *The Kalahari Environment* with Paul A. Shaw.

Pages 226–233

DR PHILIP WHITFIELD is Reader in (Associate Professor of) Parasitology at King's College, University of London. He lectures and broadcasts on many aspects of the biological sciences and has written a range of books on environmental subjects including *Jungles*, *The Hunters*, *The Animal Family*, *Rhythms of Life*, *The Longman Illustrated Animal Encyclopedia* and *The Atlas of the Living World*. He received a Christopher Award in 1990 in the "Books for Young People" category for *Can the Whales be Saved?* His research interests include the control of human parasitic diseases in developing countries and the analysis of the ecology and reproduction of parasites.

Pages 6–23; 30–53; 100–101; 140–141; 160–163; 168–173; 184–191; 204–217.